# THE GEOLOGICAL EVOLUTION OF THE BRITISH ISLES

by

T.R. OWEN M.Sc.,F.G.S.

Professor of Geology, University College of Swansea

**PERGAMON PRESS**

OXFORD · NEW YORK · TORONTO · SYDNEY · PARIS · FRANKFURT

| U.K. | Pergamon Press Ltd., Headington Hill Hall, Oxford OX3 0BW, England |
| U.S.A. | Pergamon Press Inc., Maxwell House, Fairview Park, Elmsford, New York 10523, U.S.A. |
| CANADA | Pergamon of Canada, Suite 104, 150 Consumers Road, Willowdale, Ontario M2J 1P9, Canada |
| AUSTRALIA | Pergamon Press (Aust.) Pty. Ltd., P.O. Box 544, Potts Point, N.S.W. 2011, Australia |
| FRANCE | Pergamon Press SARL, 24 rue des Ecoles, 75240 Paris, Cedex 05, France |
| FEDERAL REPUBLIC OF GERMANY | Pergamon Press GmbH, 6242 Kronberg-Taunus, Hammerweg 6, Federal Republic of Germany |

First edition 1976
Reprinted (with corrections and additions) 1978
Reprinted 1981

**Library of Congress Cataloging in Publication Data**

Owen, Thomas Richard
The geological evolution of the British Isles.
Includes bibliographies and index.
1. Geology-Great Britain. 2. Geology-Ireland.
I. Title
QE261.095        1976 554.1        75-42128
ISBN 0 08 020461 9 hardcover
ISBN 0 08 020460 0 flexicover

*In order to make this volume available as economically and as rapidly as possible the author's typescript has been reproduced in its original form. This method unfortunately has its typographical limitations but it is hoped that they in no way distract the reader.*

*Printed in Great Britain by A. Wheaton & Co. Ltd., Exeter*

R37/3

4.95

**PERGAMON INTERNATIONAL LIBRARY**
**of Science, Technology, Engineering and Social Studies**

The 1000-volume original paperback library in aid of education,
industrial training and the enjoyment of leisure

# GEOLOGICAL EVOLUTION

# OTHER TITLES OF INTEREST

ALLUM, J.A.E.

Photogeology and Regional Mapping

ANDERSON, J.G.C. and
OWEN, T.R.

The Structure of the British Isles

BROWN, D.S.W., CAMPBELL, K.S.W.
and CROOK, C.A.W.

The Geological Evolution of Australia

KEEN, M.J.

Introduction to Marine Geology

PRICE, N.J.

Fault and Joint Development in Brittle
and Semi-brittle Rock

SIMPSON, B.

Geological Maps

SIMPSON, B.

Rocks and Minerals

SPRY, A.

Metamorphic Textures

The terms of our inspection copy service apply to all the above books.
Full details of all books listed will gladly be sent upon request.

# Contents

# Preface

This is not meant to be another book on British stratigraphy for which the reader is referred to the books written by Rayner, by Wells and Kirkaldy and by Bennison and Wright. The volume "Earth History" (in two parts) by Read and Watson gives the details of World stratigraphy and the evolution of the various continents.

This present book attempts to outline, in narrative form, the geological history and evolution of the British Isles and its surrounding seas. This area has moved around, covering many degrees of latitude and longitude. It has been situated south of the Equator and on the Equator in the past. It is therefore understandable that its climate has changed considerably and that its regional setting has changed through coral seas, hot deserts, tropical swamps and even ice sheets and tundra. Ancient oceans have opened and subsequently closed. Mountain chains have risen only to be worn down to their very roots.

Geological thought has recently experienced a tremendous revolution. Palaeomagnetism and Plate Tectonics have given a "shot in the arm" to the science and there is a constant need to review opinions of the British geological story in these times of change. New information has been obtained, especially for the geology of the sea floor around Britain, and there is therefore a need to incorporate these new facts into the traditional British story which dealt largely with the evolution of those areas which to-day are land. The sea area around Britain more than doubles that of the actual land areas.

Besides the books mentioned above, further suggested reading is given after each chapter. Reference should also be made to the maps compiled by the Institute of Geological Sciences, particularly the geological maps of Britain on a scale of ten miles to the inch.

# Acknowledgements

I wish to express my sincere thanks to my colleagues in the Department of Geology, University College of Swansea, for many stimulating discussions. I must particularly single out Professor D. V. Ager for his encouragement and also Dr. T. W. Bloxam, Dr. M. Brooks and Dr. G. Kelling.

I wish also to sincerely thank Mrs. Jose Nuttall for typing my manuscript and again Mr. J. U. Edwards and Mr. G. B. Lewis for drawing the illustrations. Mrs. Alvis Smith photocopied many of the illustrations and pages of manuscript.

For the inspiration of this book I am grateful to the very many students with whom I have been associated during nearly thirty years of university teaching and also to my very many good friends in extra-mural classes. All this, however, would not have happened in the first place without the help and guidance of Professor T. Neville George who first introduced me to my subject and who was such a wonderful teacher.

The author wishes to thank many other authors for permission to adapt certain illustrations. Special thanks is recorded to W. H. Freeman & Co., the Institute of Geological Sciences, the Geological Society of London, the Association of Teachers of Geology, the University of Wales Press, the Sixth International Congress of Carboniferous Stratigraphy and Geology and the Open University for kind permission to reproduce certain illustrations.

The author also thanks Graham Trotman Dudley Publishers Ltd. for permission to adapt one of their illustrations.

CHAPTER 1

# Introduction

"The Present is the key to the Past". This is a text which has been the subject of geological "sermons" for more than a hundred years. The past is a
long past, more than 2000 million years back (there are rocks in W. Greenland
which are over 3700 million years old) and this time has been divided as
shown in fig. 1. The last 600 million years of time have been subdivided
into three eras (Palaeozoic, Mesozoic, Cenozoic) each decreasing progressively
in length. We live to-day in the Cenozoic Era. Each of these eras has
been further subdivided into periods (if one is talking about pure time) or
into systems (in terms of the rock succession deposited during that same
time). These divisions are still further subdivided. The scheme is international, though in America the Carboniferous is replaced by a (lower)
Mississippian and an (upper) Pennsylvanian division.

The very long time before the beginning of the Palaeozoic Era is generally
termed the Precambrian but it will be appreciated that this covers some
8/9 ths of the Earth's history. Because of the antiquity and the often
extremely deformed and altered character of the rocks, it is difficult to
erect divisions which can be satisfactorily applied on an international scale.
It is however recognized that a younger Proterozoic portion (base at about
2000 million years ago) can be distinguished from an older Archaean portion.

A good deal of attention has been given in recent years to these ancient
rocks. Radiometric dating of the rocks has clarified the position and is
helping to unravel the very complex early histories of these ancient areas.
It has been shown that certain stable masses, formed of these intensely
deformed rocks, are very old and form the very nuclei of the continents as
we know them to-day.

## PLATE TECTONICS

The trouble with so many geological "sermons" has been that they have
wandered away from the text (that "The Present is the key to the Past"). In
fact one could almost say that "the Past has been made the key to the
Present!". We have not, until fairly recently, known enough about the
configuration and behaviour of the present surface of the Earth. Geological
processes and phenomena have tended to be explained on a piecemeal basis.
Deep subsiding sedimentary basins or troughs (often conveniently dismissed
as "geosynclines") have been described and explained in isolation. Seas
came and went, again in isolation, and without any broader relationship.
Periods of igneous activity were never quite explained. As long as
"Continental Drift" was doubted, there was no explanation for the changes in
climate and land environments which must certainly have occurred.

Then came the beginnings of the Revolution. The detection of palaeomagnetism was a great step forward and brought in the physicist as a powerful
ally. The contributions by Runcorn were of particular importance. In 1960
Hess built on these foundations and proposed the idea of "sea-floor

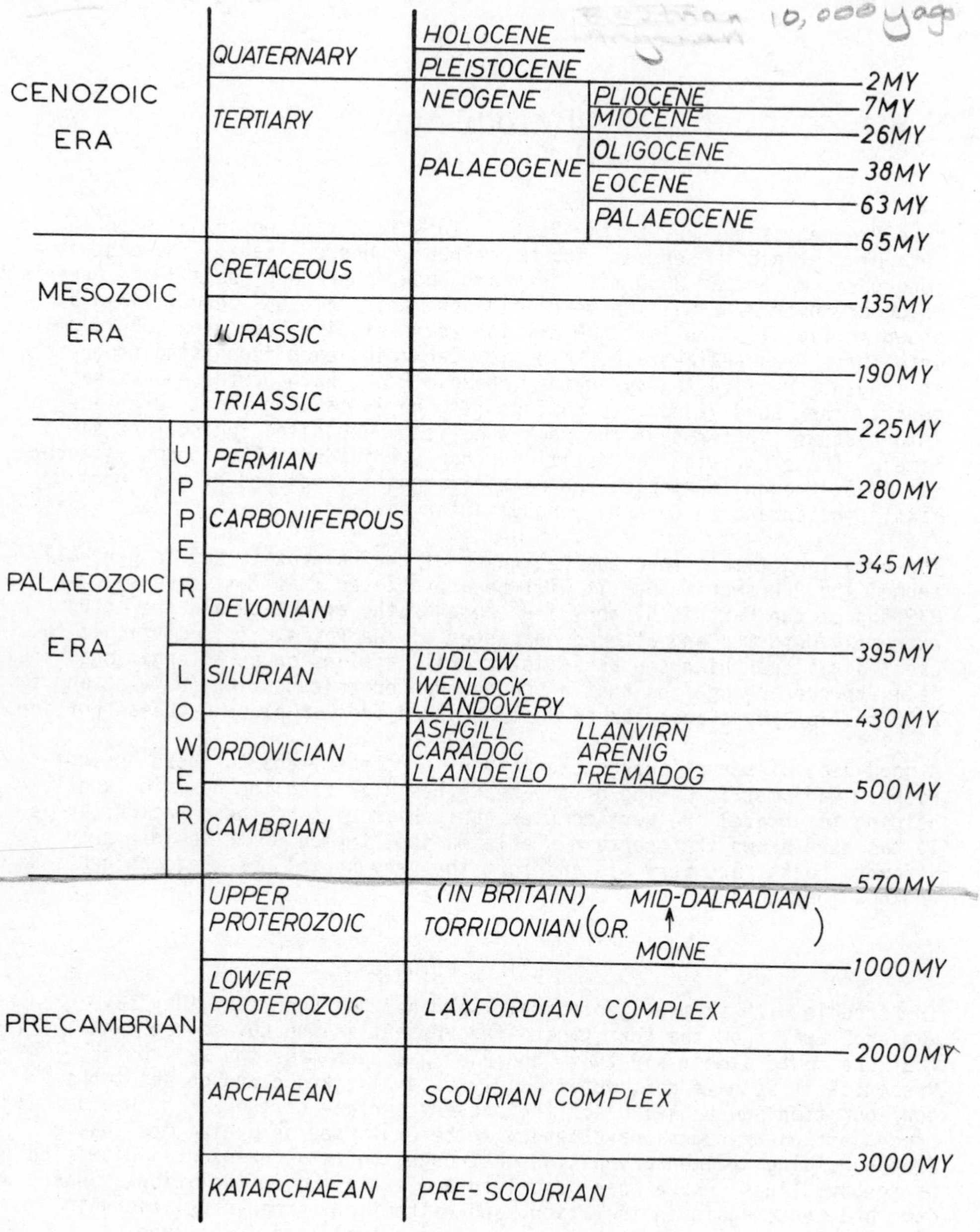

| | | | | | |
|---|---|---|---|---|---|
| CENOZOIC ERA | QUATERNARY | HOLOCENE | | | |
| | | PLEISTOCENE | | | |
| | | | | | 2MY |
| | TERTIARY | NEOGENE | PLIOCENE | | 7MY |
| | | | MIOCENE | | 26MY |
| | | PALAEOGENE | OLIGOCENE | | 38MY |
| | | | EOCENE | | 63MY |
| | | | PALAEOCENE | | 65MY |
| MESOZOIC ERA | CRETACEOUS | | | | 135MY |
| | JURASSIC | | | | 190MY |
| | TRIASSIC | | | | 225MY |
| PALAEOZOIC ERA | UPPER — PERMIAN | | | | 280MY |
| | CARBONIFEROUS | | | | 345MY |
| | DEVONIAN | | | | 395MY |
| | LOWER — SILURIAN | LUDLOW / WENLOCK / LLANDOVERY | | | 430MY |
| | ORDOVICIAN | ASHGILL / CARADOC / LLANDEILO | LLANVIRN / ARENIG / TREMADOG | | 500MY |
| | CAMBRIAN | | | | 570MY |
| PRECAMBRIAN | UPPER PROTEROZOIC | (IN BRITAIN) TORRIDONIAN (O.R. | MID-DALRADIAN ↑ MOINE ) | | 1000MY |
| | LOWER PROTEROZOIC | LAXFORDIAN COMPLEX | | | 2000MY |
| | ARCHAEAN | SCOURIAN COMPLEX | | | 3000MY |
| | KATARCHAEAN | PRE-SCOURIAN | | | |

Fig. 1. The geological time-scale.

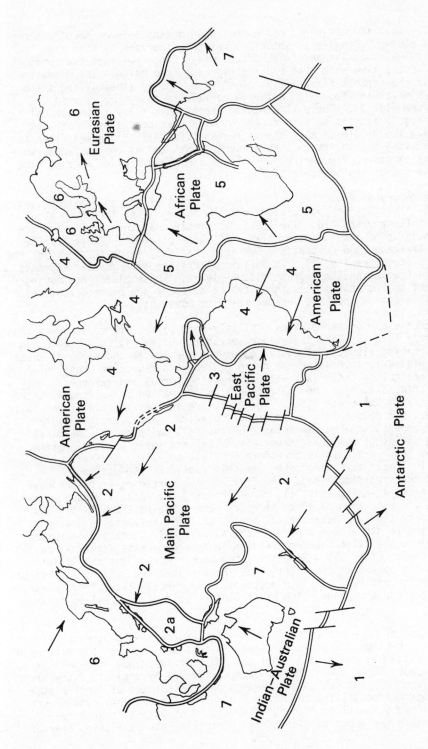

Fig. 2. The main plates of the World today (from The Origin of the Oceans by Sir Edward Bullard. Copyright©1969 by Scientific American, Inc. All rights reserved).

spreading" - that if continents had moved, then oceans between them must have opened and closed.  Equally as important was the discovery by Vine and Matthews in 1963 that linear strips of the ocean floor retained the remanent magnetism of the time when that floor strip came into being.  The magnetic polarity of the earth's field "flips over" from time to time, giving there-fore strips of ocean floor with either a normal (as today) or reverse polarity (see fig. 3).  These alternating strips make today a symmetrical ("mirror image") pattern on either side of mid-ocean crests, such as that running down the middle of the Atlantic Ocean (fig. 4).  New ocean floor is made today along the splitting line of such crusts and as the two edges move away from each other, the centre "infilling" is itself divided to each side (hence  the symmetrical pattern of fig. 3).

The modern theory of Plate Tectonics has thus eventually been developed and it must at this stage be appreciated that the theory is no more than about twelve to fifteen years old.  The thesis is simply that the Earth's surface today is made up of about 7 major plates ("jigsaw pieces").  These are the Eurasian, African, Indo-Australian, American, Pacific, Nazca (E. Pacific) and Antarctic plates (see fig. 2).  In addition there are some smaller segments such as the Arabian and Caribbean "micro"-plates.  These plates are moving with respect to one another and the rates of movement are being discovered.  The arrows in fig. 2 show the general relative movements.

Plate movements are of three kinds.  Firstly some plates are moving away from one another.  This is where "sea-floor spreading" - the making of new ocean crust - is taking place.  The distance between the two facing edges of these diverging plates is increasing and being filled with new ocean crust, of igneous (once molten) material, with alternating normal and reversed magnetism (fig. 3 a, b).  The "spreading line" is marked by volcanoes, often rising high from the ocean floor.  The extrusive igneous material is of basic composition, the deeper infillings are of dolerite.  The extrusive magma frequently congeals on the new ocean floor as "pillow lavas".  The basic in-filling strips of new ocean crust yield, besides their palaeomagnetism, evidence of their age.  It is therefore possible to assess the rate of spreading (or "half-spreading" when one only considers the distance of a strip from the spreading line).  The East Pacific spreading is faster than that in the mid-Atlantic (fig. 4).  Greenland and the Rockall Plateau are separating today at about 1.18 cm in every year.  The rate was faster in early Tertiary times.  Nevertheless the North American plate is moving westwards the length of one's body in one's lifetime!  Variation in spread-ing rate along the plate boundary results in numerous cross fractures called transform faults (fig. 3).  Note the large number across the mid-Atlantic and E. Pacific spreading lines (fig. 4).  Here plates are sliding against one another, the second type of plate movement.  Large-scale lateral or trans-current fractures are known.  The great Alpine Fault of New Zealand is an example.

The third type of movement is where two plates approach each other.  Before describing this situation it is necessary to consider the form and make-up of plates.  A plate may comprise a continent and some adjacent, or even sur-rounding, ocean.  Some plates may be covered largely by ocean.  Every plate is about 70 to 80km thick and is made up of crust on top of upper mantle (the two together comprising the "lithosphere").  The boundary between crust and mantle is the so-called "Moho".  The base of a plate is therefore situated within the upper portion of the Earth's mantle at a thin layer of

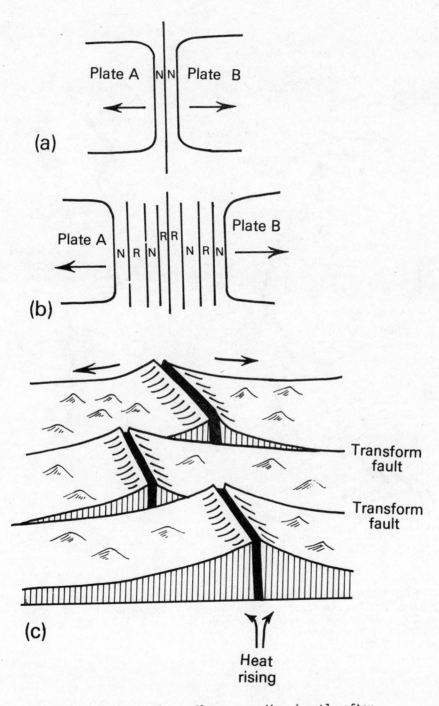

Fig. 3.  Aspects of sea-floor spreading (partly after Menard, 1969).

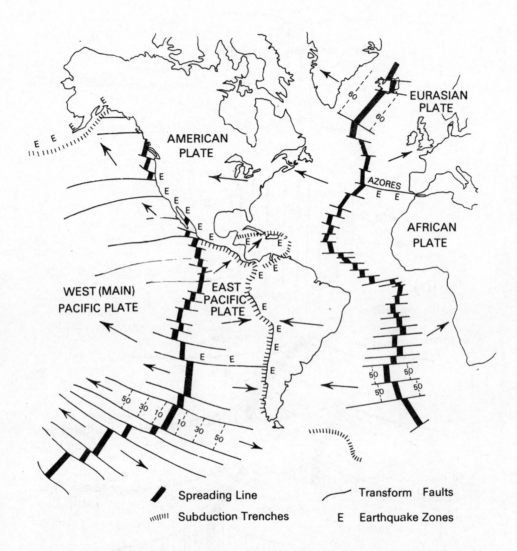

Fig. 4. Sea-floor spreading and subduction zones in the Atlantic and East Pacific oceans (from Sea-Floor Spreading by J. R. Heirtzler. Copyright © 1968 by Scientific American, Inc. All rights reserved).

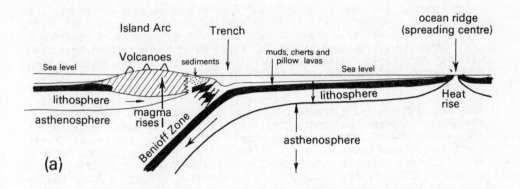

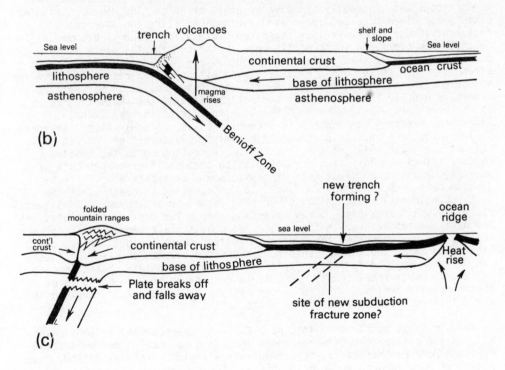

Fig. 5.  Types of plate boundaries (after several sources).

melting known as the "Low Velocity Zone" - so-called because earthquake shear
waves decrease temporarily within it before increasing once more down to the
core-mantle boundary (at a depth of 2900 km).  Under continents, continental
crust (higher in silica content) is thick (35 km on average) but under the
oceans the more basic crust is only 8 km in thickness so that there is here
a much greater thickness of upper mantle beneath the thin crust (fig. 5).
That ocean crust has a marked layering with a thin sedimentary smear of muds
and cherts passing progressively downwards into pillowed basalts, dolerite
intrusives, gabbros and ultrabasics such as peridotite.  It follows that
ocean plate is denser than continental plate and here we have the secret of
plate behaviour when segments approach one another.  Hess was the first to
suggest that if new crust was added at spreading lines then in some other
areas crust must disappear and be lost.  Otherwise the Earth's surface would
be expanding and this is not thought to be the case.  Crust is lost when two
plates move towards one another.  If ocean crust approaches continental crust
then the former, being denser, will slide down under the latter.  This is
happening today at the western edge of South America (fig. 4). The E.Pacific
Plate is sliding down under the westward moving leading edge of the (S)
American Plate (fig. 5b).  The slide boundary is called a "subduction" or
Benioff zone and is usually at about an angle of 45°.  The downward sliding
ocean plate gives rise to deep-seated earthquakes along the American
continental edge.  Moreover, as the ocean plate slides down it becomes
melted.  The overlying upper mantle beneath the continental plate is also
partially melted so that molten magma sources are provided, giving rise to
ascending magma (often of an andesitic composition) which results in numerous
volcanic outbursts along the continental edge (fig. 5b).  A marked deep
trench is usually found above the surface position of the subduction zone.
As the ocean plate slides beneath the leading edge of the continent,
frictional scraping can occur with irregular masses of ocean floor sediment,
pillow lavas, even deeper ocean crust becoming "plastered" or "obducted"
against the inner walls of the trench, i.e. accreting on to the leading
continental edge.  This is a fortunate thing because it preserves bits of
ocean floor and ocean crust that otherwise would disappear down the subduction belt.  Ocean crust is doomed to destruction, no ocean plate can
last.  Continental crust on the other hand cannot be destroyed. A continental
plate may split anew along a new spreading line.  Two continental plates can
approach one another (as in fig. 5c).  Here both plate edges are of light
density and no subduction takes place.  Instead there is continental collision and thick sedimentary sequences deposited between the two continents
can be intensively deformed and uplifted into high mountain chains (adding
in fact to continental crust).  Hot magma can deeply invade the roots of
these mountain chains.  This is partly because any ocean plate involved in
the collision breaks off below and melts in the mantle.

In fig. 5c it is possible to imagine a further repercussion, at some point
between the colliding continents and a spreading centre.  A new subduction
fracture zone can be set up, this time entirely within an ocean plate. Ocean
plate can here subduct under ocean plate.  The descending slab melts and the
rising magma gives rise to lines of volcanic islands.  This is an island arc
(fig. 5a), the magma again being of a calc-alkaline character.  An island
arc can then form, of course, on the (near) seaward side of a continent.
Japan and the Aleutian Islands are examples.  Deep trenches occur on their
oceanward sides and severe deep-seated earthquakes often occur, sited on
the Benioff zone.  Island arcs can fringe a continental edge with even (thin)
continental crust behind them.  Obduction of portions of ocean crust can be

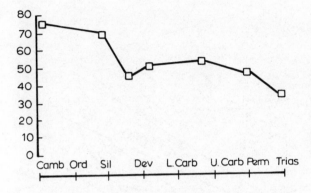

Fig. 6. Present latitudes of past equators (after Schove, 1964).

Fig. 7(a). Underground limits of buried Mesozoic divisions (after George, 1962). Key: Coarse dots - Triassic; inclined shading - L.Jurassic; Blank - M. & U. Jurassic; Fine dots - L.Cretaceous; Horizontal shading - U.Cretaceous.

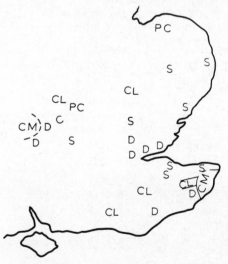

Fig. 7(b). Sub-Mesozoic geology of S. E. England (after George, 1962). Key: P.C., Precambrian; C., Cambrian; S., Silurian; D., Devonian; C. L., Carb. Limestone; C. M., Coal Measures.

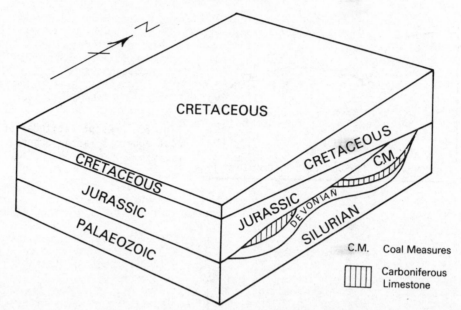

Fig. 8.   Underground structure of E. Kent.

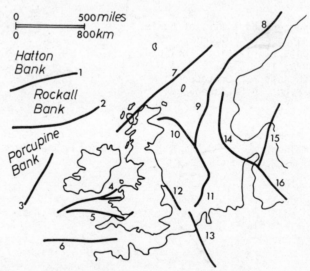

Fig. 9.  Main troughs around the British Isles (after
Naylor and Mounteney, 1975).  Key:  1. Rockall-Hatton
Trough.  2. Rockall Trough.  3. Porcupine Trough.
4. N.Celtic Sea Trough.  5. S.Celtic Sea Trough.
6. W. Approaches Trough.  7. W. Shetland-Minch Trough.
8. Voring Basin.  9. N. North Sea Trough.  10. Forties
Trough.  11. N. Netherlands Trough.  12. West Sole
Trough.  13. Lower Rhine Trough.  14. W. Norway Trough.
15. Oslo Rift.  16. Danish-Polish Trough.

smeared against the leading edge of the island arc.  Abnormally high pressure conditions combined with low temperature near the upper part of the
Benioff zone can result in a special metamorphic environment producing a
characteristic suite of rocks such as blue glaucophane schists and serpentinites.  These can be found in a faulted maze of obducted cherts, muds,
pillow lavas and basic intrusives.  Such an irregular assemblage is known as
an "ophiolite suite" and is associated with modern zones of subduction.
They make an excellent case for the application of our text "The Present is
the key to the Past".  Ancient ophiolite assemblages point to the position of
past subduction zones and ancient plate sutures.  Such "fossilised" subduction belts include the Ordovician suites at Girvan and the even earlier
(late) Precambrian assemblage in Anglesey.

It was considered necessary, in this Introduction, to deal with the main
elements of Plate Tectonics in some detail.  The publications of, for
example, Dewey, and of Ziegler and McKerrow have stimulated a new look at the
stratigraphy of areas such as Britain.  Moreover, the new global tectonics
have shown the close relationship between the various geological processes -
sedimentation, igneous activity, mountain building, earthquakes, metamorphism, rifting.  Furthermore, it is now possible to unravel the result of
palaeomagnetic fixing in terms of plate positions in the past.  Note, for
example, the high latitude positions today of points that in the Palaeozoic
were on the Equator (fig. 6).  There is a hint here of a northward drift,
overall, of parts of the Earth's segments.


## BENEATH THE LAND, BENEATH THE SEAS

Reconstruction of past geological history must always suffer from incomplete
information and every palaeogeographic map must be treated with some sus-
picion.  Missing sequences particularly present problems.  Such strata could
once have been deposited and then subsequently removed.  There is always a
tendency to place limits to deposition only just beyond their preserved
limits today.  Reconstructions of Chalk palaeogeography even now cannot
resist the temptation to put an island over Snowdonia or to limit Chalk
deposition in Southern Ireland to the discovered patch near Killarney.  The
lesson of the Mochras Borehole has, one hopes, been learnt.  Here, in the
heart of a permanent land area of Jurassic reconstruction, was discovered the
thickest (1300 m) marine Lias sequence in the British area.

It is also important in palaeogeographic reconstruction to try to determine
the geology at depth.  The discovery of granites beneath the rigid Alston and
Askrigg blocks adds to our knowledge of late Silurian igneous emplacement and
helps to understand the peculiar rigidity of that north Pennine area in
Carboniferous times.  There are further pointers now, in the case of K-Ar
determinations of the Wensleydale Granite, of a later (Upper Carboniferous)
thermal event, which can now be related perhaps to early (belated) attempts
at crustal stretching (spreading?) in northern Britain.

Borehole information has now amounted to a formidable pile in Britain and
this helps reconstruction considerably.  The two samples quoted from the work
of George (fig. 7) can be sited.  Fig. 7a reconstructs the eastward overlap
of successively younger Mesozoic units against the pre-Mesozoic basement and
thereby outlines the western edge of the East Anglian Ridge.  The surface
geology of that area today shows only Chalk and Cenozoics.  The actual age

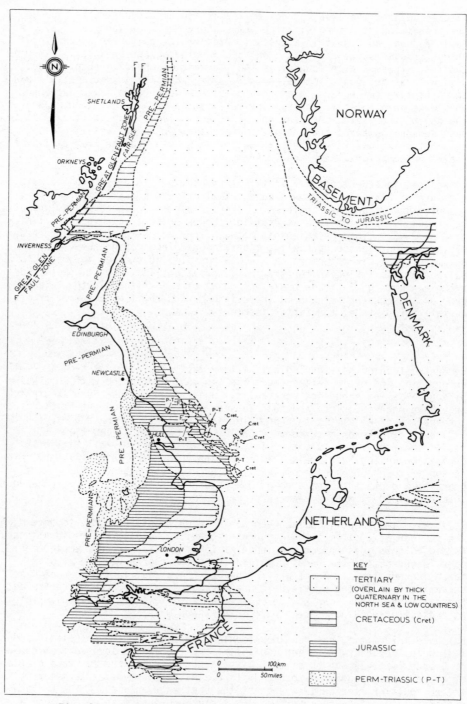

Fig. 10. The Post-Carboniferous geology of the North
Sea and adjacent areas (based on the I.G.S. map of 1972).

of the surface of the buried basement (fig. 7b) is also of interest.  Much
of that foundation shows Silurian, Devonian and Lower Carboniferous, but at
North Creake (in Norfolk) Triassic is directly underlain by Precambrian
(Uriconian or Charnian tuffs).  The presence of those Precambrian volcanics
extends still further the extensive vulcanicity found from Pembrokeshire to
the Welsh Borderland and Central England.  Again the Devonian areas point to
unexpected marine incursions into what was previously considered to be a
Devonian land area.  The buried Kent Coalfield (fig. 8) was of course dis-
covered many years ago.  Its Burford rival is a much more recent discovery.
The presence of Pennant-type molasse in all these coalfields from S.Wales to
Kent is an important contribution to Upper Carboniferous reconstruction (see
Chapter 5).

As long as our knowledge of sea-floor geology was not available, reconstructing the geological history of the British area had to be incomplete and
speculative.  The water areas of the North Sea, Irish Sea and English Channel
together make up more than the land areas of the British Isles. The distance
from Northumberland to S.W. Norway is as much as that from Hampshire to the
Forth.  Yet until very recently we did not know about the thick Cenozoic
sequences in the troughs of the North Sea, or of the much greater extent of
Permian evaporites there. Many important sedimentary troughs or grabens have
now been located beneath the waters around Britain (fig. 9 is adapted from
the recent excellent book by Naylor and Mounteney).  As a result one can
expand the areas of known sedimentation in Permian, Mesozoic and Cenozoic
times.  Moreover, these troughs help to unravel the evolution of the North
Atlantic openings and, of course, will considerably help our economy as even
more oil and natural gas reserves are detected within them.  The geological
map of the North Sea shown in fig. 10 was just not available fifteen years
ago.  What is also important is we now know the deeper geology and structure
underneath that vast sea floor expanse of Cenozoic sediments.  Knowledge of
the complex geology and structure of the South Irish and Celtic seas (fig. 73)
and of the English Channel (fig. 79) considerably aid one's reconstruction of
the evolution of those areas.  One important fact that has emerged is the
widespread unconformity at the base of the Upper Cretaceous.  This again
relates to an important phase in the openings of the North Atlantic Ocean.


Suggested Further Reading

Bennison, G.M. & Wright, A.E. 1969.  The Geological History of the British
  Isles.  Edward Arnold. London.
Bullard, Sir Edward. 1969.  The Origin of the Oceans.  In The Ocean.
  Scientific American.  W.H. Freeman & Co.  San Francisco.
Dietz, R.S. & Holden, J.C. 1972.  The Breakup of Pangaea.  In Continents
  Adrift.  Scientific American.  W.H. Freeman & Co.  San Francisco.
Heirtzler, J.R. 1972.  Sea Floor Spreading.  In Continents Adrift.
  Scientific American.  W. H. Freeman & Co. San Francisco.
Menard, H.W. 1969.  The Deep-Ocean Floor.  In The Ocean.  Scientific
  American.  W.H. Freeman & Co. San Francisco.
Naylor, D. & Mounteney, S.N. 1975. Geology of the North-West European
  Continental Shelf. Vol. 1.  Graham Trotman Dudley Publishers Ltd. London.
Oxburgh, E.R. 1974.  The Plain Man's Guide to Plate Tectonics.
  Proc. Geol. Ass., 85, 299.
Rayner, D.H. 1967.  The Stratigraphy of the British Isles.   Cambridge
  University Press.  Cambridge.

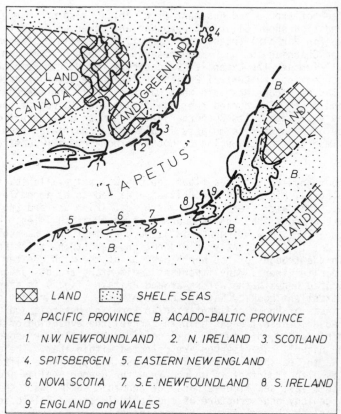

Fig. 11. Cambrian palaeogeography and faunal provinces
(after Cowie, 1974 and Marshall Kay, 1969).

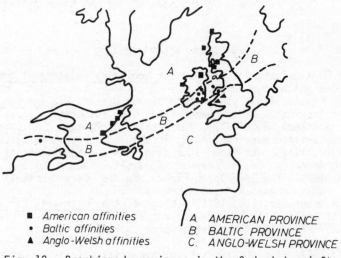

Fig. 12. Brachiopod provinces in the Ordovician (after
Williams, A. 1969).

CHAPTER 2

# "Iapetus"--the First Atlantic Ocean

The notion of a "Proto-Atlantic" Ocean which had opened in the Lower
Palaeozoic and closed again is mainly associated with J. T. Wilson. The
idea has grown in strength with important contributions by Dewey, Kay, and by
McKerrow and Ziegler. In brief, it is believed that the Lower Palaeozoic (and
possibly also the later Proterozoic) sediments of Britain accumulated on the
two margins of an ocean that separated Scotland and N.Ireland on the one side
(the northern side) and England, Wales and S.Ireland on the other. These
northern and southern halves of Britain were then a long distance from each
other in Lower Palaeozoic times. The ocean may have been narrow to begin
with, in late Precambrian times, reached its widest extent in the Cambrian
period and then began to narrow in Ordovician and especially in Silurian
times. Not only parts of Britain may have been thus separated. Newfoundland
was similarly split into two widely separated portions. The same Proto-
Atlantic Ocean (or "Iapetus" as it has since been known) separated Greenland
from Scandinavia and the St. Lawrence area from Nova Scotia and the New
England States (figs. 11 and 14). By the end of the Silurian the ocean had
closed (at least in its British region - it was not to finally close in the
N.American portion until slightly later) and the two halves of Britain were
wedged together. There is some controversy about the precise position of
the "join" but it could be somewhere in the South of Scotland - perhaps about
the position of the Solway Firth (Gretna Green may well have a geological
significance in terms of the crustal "marriage" of northern and southern
Britain). Others think the Southern Uplands Fault could represent the
position of the welding.

The Lower Palaeozoic of Britain has always posed interesting problems. Its
great rock thicknesses in certain parts of Britain (and in North America)
gave rise to the idea of a geosyncline - a great downsag of the earth's crust
which thereby allowed the accumulation of thousands of feet of sediments (and
volcanics). Ideally, the Lower Palaeozoic Geosyncline (in Britain) had its
northern shoreline off N.W. Scotland and its southern shoreline fluctuating
back and fore somewhere over central and southern England. One identified
nearshore deposits with brachiopod-trilobite-coral assemblages and more open
water dark muds with graptolite assemblages. The great Ordovician volcanic
outbursts were believed to be of submarine origin. This simple model for the
Lower Palaeozoic gradually gave way to more complex pictures. Lateral
variations in thickness and lithologies necessitated postulating a number of
basins of more important accumulations. Moreover, marked unconformities
separated each of the Lower Palaeozoic systems in Britain pointing to marked
contractions of the geosyncline and its various units. A further complication
was the gradual recognition of ignimbritic, welded volcanics which were
probably of subareal, rather than submarine, origin. One enlightened aspect
of these geosynclinal reconstructions was the realisation (relatively early
in the Lower Palaeozoic Geosyncline story) of distinct faunas along the
northern and southern margins of the trough. The trilobite and brachiopod
faunas of these two margins differed from one another. Theories to account
for the differences ranged from too deep waters to poisonous waters or even
separating land barriers. These faunal differences were, however, a pointer
in the right direction, as will be seen presently.

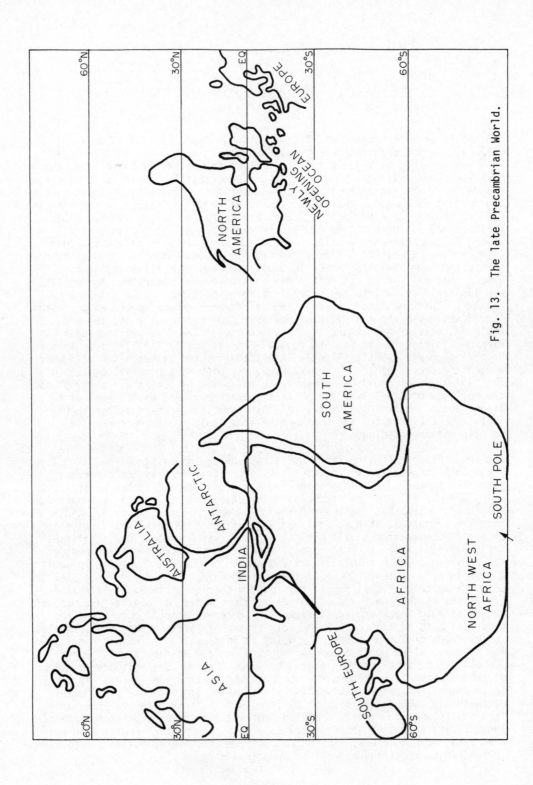

Fig. 13.   The late Precambrian World.

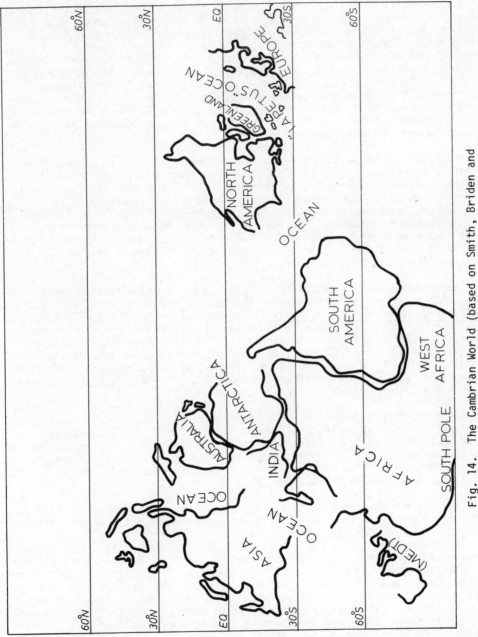

Fig. 14. The Cambrian World (based on Smith, Briden and Drewry).

According to the geosynclinal idea, then, the sediments accumulated on a
(variably) subsiding crustal floor which was ultimately buckled into a folded
orogen.  The folded floor was however still there, though of course it could
have been vastly altered by the metamorphic processes involved in the base
of a great orogen.  The differences in the new theory must therefore be
appreciated.  In the geosynclinal theory, two "shoreline" margins X and Y
would have had their greatest separation distance at the commencement of the
geosynclinal cycle.  That X-Y distance could only then get shorter with time
and with crumpling.  In the plate tectonic theory however, the two margins
X and Y could get increasingly further apart due to sea-floor spreading
processes acting in an area between them. Ultimately with subduction gaining
sway over spreading, the margins X and Y could again approach one another and
from here onwards the original X-Y distance could be appreciably lessened by
crumpling and therefore shortening.

At this point it is worth considering the general position of plates in the
Lower Palaeozoic world.  Figs. 13 and 14 show the probable positions of our
modern land areas in late Precambrian and in late Cambrian times.  It can be
seen that there was much more land in the southern hemisphere with very
little land indeed north of 30°N.  Notice that Australia and S.E. Asia were
in the northern hemisphere!  Britain was south of the equator but mainly
within or near tropical latitudes.  The North Pole was out in the Pacific and
the South Pole was in N.W. Africa.  The Equator ran from N.Greenland to
Spitzbergen in the late Cambrian.

Trilobite faunas delineate the two margins of the Proto-Atlantic Ocean
(fig. 11) in Lower Cambrian times.  The Pacific Province with Olenellus,
Bathynotus, Nevadia, Bonnia, Protypus etc. characterized the northern margins
of "Iapetus" whilst the Acado-Baltic Province on the southern margins had
Callavia, Holmia, Kjerulfia and Strenuella.  In the Lower Ordovician the
northern margins were characterized by a Bathyurid fauna whilst the southern
margins yielded a Selenopeltid fauna.  The latter is believed to have lived
in somewhat cooler waters than the northern fauna.  The Baltic margins however had an Asaphid fauna at first.

Distinct brachiopod provinces are also found in the early Ordovician with a
Scoto-Appalachian assemblage in N.W. Scotland and N.W. Ireland and quite
different faunas (Baltic, Celtic and Anglo-French) in S.E. Ireland, Wales and
the Welsh Borderland.  By Upper Ordovician times however, the differences
between these trilobite and brachiopod realms have disappeared and an essentially cosmopolitan fauna had emerged.

Faunal differences like this raise the question of the width of the Proto-
Atlantic Ocean.  Was it as wide as the modern North Atlantic or was it a much
narrower seaway?  Has the degree of early Palaeozoic spreading been
exaggerated?  Some say it has.  On the basis of Palaeomagnetic results for
Eocambrian tillites, Tarling (1974) finds it difficult to maintain a width
much greater than 500 km for the Proto-Atlantic, in contrast to a much wider
Tethys Ocean, perhaps as wide as 3500 km.  Briden and Morris thought that by
the Ordovician the Proto-Atlantic Ocean was already small.  Ueno sees two
possibilities.  Either (1) the Caledonian Orogeny was not the site of a wide
early Palaeozoic ocean and the bounding cratons were displaced sinistrally
(this is to fit palaeomagnetic results) or (2) there was a wide ocean and the
bounding shields returned to about their original positions.  Cramer and Diez,
from studies of acritarchs, think that the Ordovician Proto-Atlantic was
considerably narrower than the present Atlantic.  In contrast, Dunning in

1972 thought that current speculation envisaged an ocean 2000 km wide in the Cambrian, but closing to a smaller distance in later Lower Palaeozoic times. Strong and others, on the basis of active subduction zones over appreciable distances, saw the Proto-Atlantic as a rather large ocean basin. The problem is a difficult one, in that one is trying to reconstruct widths (that would in any case have varied from time to time and from one portion of the trough to another) over an area that was subsequently deformed (at different times) and which subsequently closed.   It must also be remembered that whereas palaeomagnetic results give palaeolatitudes, they do not indicate palaeo-longitude and varying interpretations of the latter can give widely varying widths.  The majority view is that the Proto-Atlantic may have reached a width of at least 2000 km in Cambrian times but with greater subductive activity from late Cambrian times onwards this width may never again have been reached - at least in the British area.  From late Cambrian times on-wards, the British portion of the ocean probably began to resemble the West and S.W. Pacific of today with its numerous island arcs and back arc seas.

## EARLY STAGES OF THE OCEAN

Resting with marked unconformity (depicting a very irregular early relief) on the metamorphosed Lewisian basement of N.W. Scotland is the Torridonian, a succession (up to 7,000 m thick in places) of predominantly red or chocolate arkosic grits, sandstones, flags and shales with conglomeratic layers.  This Torridonian succession has been subdivided into three formations.  The lowest (Diabaig) is overlapped northwards by the middle (Applecross) formation in the extreme northwest of the Scottish mainland.  The upper (Aultbea) form-ation disappears in places due to slight flexuring of the Torridonian beneath the Cambrian overstep.  In Islay and Colonsay, 2000 m of turbidite greywackes underlie the Diabaig sandstones and rest unconformably on the Lewisian base-ment.  These greywackes were transformed northwards.  The three main Torridonian units are however probably of fluviatile origin.  Williams (1969) has suggested that in Western Ross and Sutherland, the Applecross Formation includes at least two great alluvial fans (fig. 16).  These fans were pushed out southeastwards by rivers emerging from mountain gorges somewhere in the vicinity of the present Outer Hebrides.  From studies of clasts in the Apple-cross Formation, Allen, Sutton and Watson have confirmed their derivation from a Lewisian basement landmass to the northwest (fig. 17).  Radiometric ages of tourmaline-quartz pebbles taken along the main outcrop fit the pattern of the Precambrian basement of Greenland.

The great Moine Thrust of N.W. Scotland separates the non-metamorphosed Torridonian Group from the metamorphic Moine Series (or Moinian). This series reaches a thickness of almost 7000 m in places and dominates the surface geology of a huge tract of the Scottish Highlands, from the great thrust zone to the western edge of the Grampians.  Radiometric datings (1000-800 m.y.) suggest that the Torridonian and Moine could be the lateral equivalents of one another.  Before complex folding and Caledonian metamorphism the Moine rocks were a great thickness of badly sorted sandstones with muds, the whole probably of deltaic origin.  Current bedding directions indicate a northward to northeastward transport parallel to the margin of the Moine basin (fig.16). Thus the picture for N.W. Britain 1000 to 800 million years ago emerges as (a) high mountains to the northwest, made of Lewisian (probably supra-crustal?) rocks, (b) irregular gorge-like tracts and with very irregular relief, leading seawards to a deltaic coastline with a developing trough. The Moine sediments may thus be seen as the first shelf-edge deposits of the early "Iapetus" (fig. 15).  That such a situation was already an unstable one

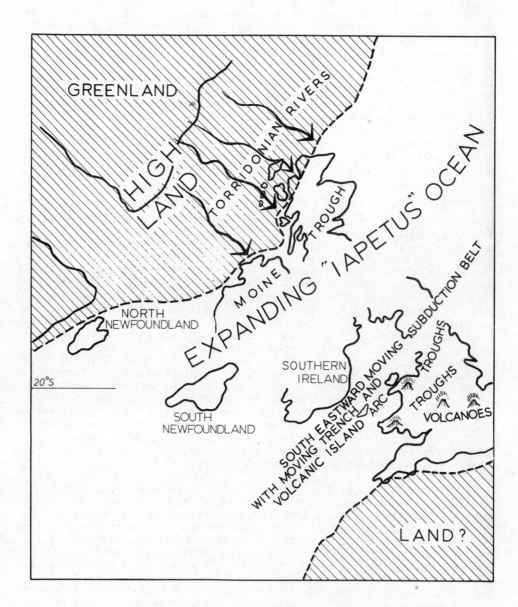

Fig. 15.  "Iapetus" in late Precambrian times.

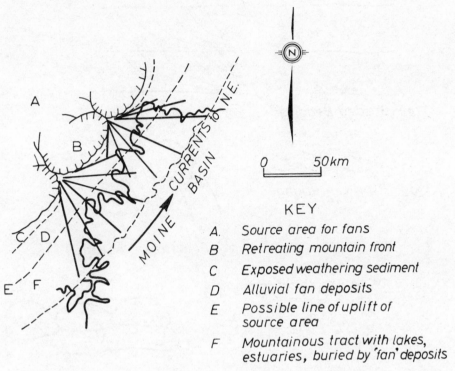

KEY

A.   Source area for fans
B.   Retreating mountain front
C.   Exposed weathering sediment
D.   Alluvial fan deposits
E.   Possible line of uplift of
     source area
F.   Mountainous tract with lakes,
     estuaries, buried by 'fan' deposits

Fig. 16.  Torridonian alluvial fans (after Williams,
G.E. 1969).

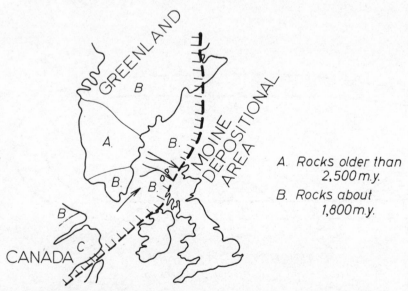

A. Rocks older than
    2,500 m.y.
B. Rocks about
    1,800 m.y.

Fig. 17.  Source geology for Torridonian (after Allen,
Sutton and Watson, 1974).

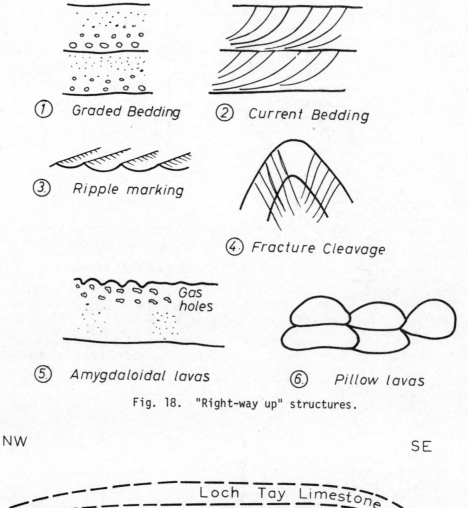

①   Graded Bedding      ②   Current Bedding

③   Ripple marking

④  Fracture Cleavage

Gas holes

⑤   Amygdaloidal lavas      ⑥   Pillow lavas

Fig. 18.   "Right-way up" structures.

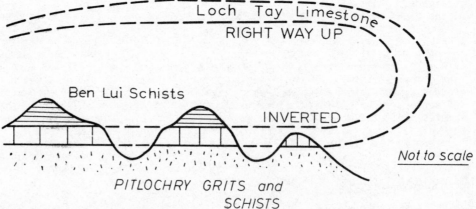

NW                                                  SE

Loch Tay Limestone
RIGHT WAY UP

Ben Lui Schists

INVERTED

Not to scale

PITLOCHRY GRITS and
SCHISTS

Fig. 19.  Diagrammatic sketch of the inverted succession
in the Loch Tay "Flat Belt" (near Balquhidder).

is shown by an early Caledonian folding and metamorphic imprint (dated at about 750 m.y.) in the westernmost (and probably deepest buried) Moine.

The Moinian is overlain by the thick Dalradian Supergroup, rocks which dominate a broad area across the Scottish Highlands from the Grampians through Perthshire to Kintyre and on into Northern Ireland. The Dalradian reaches a thickness of over 8000 m and was probably in the main deposited in areas east and southeast of the Moine region. It is likely that the lowest Dalradian might even be the lateral equivalent of upper Moine sediments. The Dalradian probably spans an appreciable time. The Lower Dalradian is Precambrian and the base of the Cambrian probably falls some-where within the Middle division of the supergroup. Well-marked tillite horizons at the base of this middle division are perhaps a convenient marker for the base of the Scottish (and Irish) Palaeozoic though this glacial episode (which was almost world-wide) is generally considered to be of Eocambrian (or Infracambrian) age. The Upper Dalradian has yielded (at Callander) fossils indicating a low Middle Cambrian horizon. While the Lower Dalradian almost continues a Moine-type deposition (though with some calcareous sediments now), the middle and upper division represent much more turbidite-type sediments, whilst volcanics (with pillow-structure) and ashes also make their appearance in places in the Upper Dalradian.

The Dalradian rocks are much folded and metamorphosed though fortunately the metamorphism is variable in intensity and it is often possible to make out sedimentary and other structures which give the "right-way-up" or "younging" direction (fig. 18). In these ways it has been possible to demonstrate the great nappe folds and slides of the Dalradian tract. One notable example is the great Loch Tay nappe (fig. 19). The folding and metamorphic history of the Dalradian is (like the Moine) very complex. Numerous phases of folding affected the Dalradian probably before the end of Cambrian times, certainly by mid-Ordovician times, and the mass was metamorphosed by that latter time too. Radiometric dates for the Dalradian concentrate around 470-440 m.y. but are somewhat younger (430-400) for the Moine pile. Presumably this difference reflects the earlier cooling of the high crustal position of the Dalradian layer.

So far, the northern margins only of the early "Iapetus" have been considered. What of the southern side? It is difficult to assess the ages of the oldest Precambrian rocks exposed in the numerous small and scattered inliers of England, Wales and S.E. Ireland. The oldest of these inliers may possibly be that at Rosslare in Ireland where Lewisian-age rocks appear to be present (on recent radiometric evidence, see chapter 3). Radiometric dates on Uriconian volcanics from Shropshire give a minimum age of 677-632 m.y. (the dates have been interpreted as post-formational). On palaeomagnetic grounds comparisons have been made between the youngest (Western) Longmyndian sedimentary rocks and the Upper Torridonian (dated at 805 m.y. for some shale samples). If the Uriconian volcanics precede the thick Longmyndian sedimen-tary pile, then that Shropshire vulcanicity would be of much earlier Torridonian (or Moinian) age. The dating and correlation of Uriconian vulcanicity is important since it has a bearing on the time of deposition of the overlying Longmyndian and also on the age of the Charnwood Forest vulcanicity (with its widespread remnants (underground) across to even Norfolk). In support of an Upper Longmyndian-Upper Torridonian correlation is the presence of "Uriconian"-type clasts in the Torridonian. On the other hand, there is a belief (Dewey, Thorpe, Shackleton, Baker) that Longmyndian sedimentation took place at about the same time as the thick eugeosynclinal

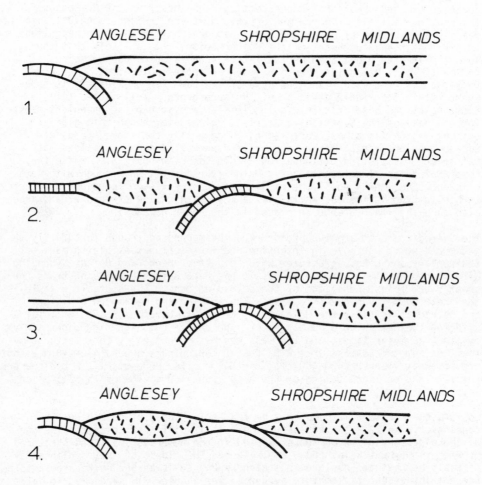

Fig. 20. Alternative plate movements for late Precambrian times, Anglesey to Shropshire (after Thorpe, 1974). Striped shading - ocean crust; broken shading - continental crust.

Monian sequence of Anglesey, both to be deformed in very late Precambrian times. One difficulty here, of course, is to say when Monian deposition began, in other words, how far back into Proterozoic times it goes. To complicate the discussion even further, recent Rb-Sr isotope data on the Stretton Series (Eastern Longmyndian), west of the Church Stretton Fault, suggest (Bath, 1974) a much younger age (maximum 600 m.y.) for this lower portion of the Longmyndian sedimentary pile. This would hardly make the Longmyndian even Precambrian! Similar doubts have been cast recently for the Ingletonian rocks of the N.W. Pennines. On the basis of these new dates, Bath rules out correlation of the Longmyndian with the Torridonian and points out that the dates for the Coedana Granite (intruded at a late stage into the folded and metamorphosed Monian of Anglesey) of 609-614 m.y. make the Monian older than the Longmyndian. Toghill (1975) believes however that these younger dates from the Longmyndian indicate a metamorphic episode in the Cambrian rather than the true age of the (affected) Longmyndian and holds to the correlation of Western Longmyndian with Upper Torridonian (at about 800 m.y.). Correlation problems affect other Precambrian inliers also, especially the metamorphic complex of the Malvern Hills. Thorpe (1974) believes that a metamorphic basement is present at widespread localities southeast of Anglesey - as the Rushton Schists of Shropshire, the Malvern Gneiss, the Primrose Hill rocks of the Wrekin. Thorpe also does not hold that the Warren House volcanics in the Malverns are necessarily younger than the Malvernian. Precambrian intrusives have also to be dated and correlated. Those in Leicestershire have recently been dated and indicate a variety of dates, 552, 433 and 311 m.y. (Cribb, 1975). The oldest of these represent intrusive activity at close to the Precambrian-Cambrian boundary. The intrusions of South Pembrokeshire could be of about this age - or could correlate with an even older porphyroid date (684 m.y.) from Leicestershire.

It is obviously difficult (and, some might say, even meaningless) to attempt to outline the late Precambrian history of this southern margin of "Iapetus" when one cannot agree on detailed regional correlations. On the other hand it can be agreed that during the later portion of Proterozoic time there occurred, in England and Wales, volcanic episodes often of calc-alkaline type but with an abundance of acid volcanics. These indicate eruption on to continental destructive plate margins rather than island arcs (Thorpe, 1974). The volcanic episodes may not correlate from place to place (the Uriconian vulcanicity being perhaps somewhat earlier than that of Pembrokeshire or even Leicestershire). Intrusive activity mostly post-dated these volcanics. The Uriconian vulcanicity continued sporadically into earlier Longmyndian times in Shropshire but then molasse-type sedimentation took over, accumulating a nearly 5000 m thick pile. This thick Longmyndian over Uriconian wedge was subsequently massively folded into an overfolded deep synclinal structure, the volcanics now forming two upstanding flanks. Cambrian rocks rest unconformably on Uriconian in the Wrekin so the deep folding in Shropshire preceded Lower Cambrian deposition.

One can go further and see major events taking place in late Proterozoic times in N.W. Wales. The Monian System of Anglesey and Caernarvonshire comprises over 10,000 m of sedimentary and volcanic rocks. The lower part of this eugeosynclinal sequence consists (according to Wood, 1974) of flysch sediments, whereas the upper part contains limestones, arenites, cherts and basic pillow lavas with a great sedimentary slide - "the melange" of Edward Greenly - occurring over a wide area. The lower Monian contains ultramafic and mafic intrusions which are serpentinized and carbonated. Late Precambrian folding was accompanied by metamorphism of very variable intensity over

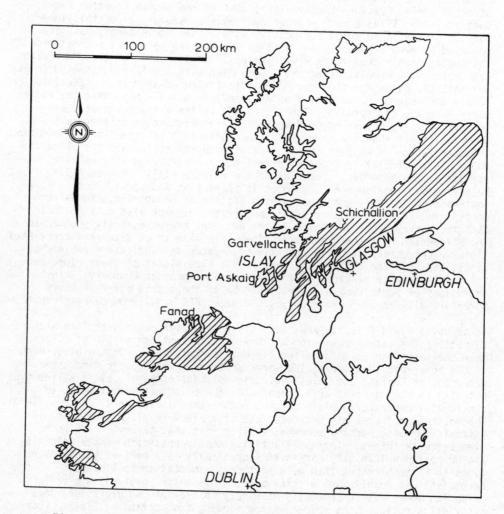

Fig. 21. Localities of the Port Askaig Tillite in Scotland and Ireland (Spencer, 1971, fig. 1). Shaded areas indicate the main Dalradian outcrops.

short distances, so that almost unmetamorphosed rocks pass into migmatites. Blueschist metamorphism proceeded in relation to major zones of contemporaneous shearing (slides) in eastern Anglesey (the famous glaucophane schists exposed near the Menai Straits). The region is interpreted as a subduction zone with an ocean plate being destroyed beneath what is now the Welsh mainland. The Arvonian ignimbrites, which span the Precambrian-Cambrian time boundary in Caernarvonshire, are interpreted by Wood as recycled upper crustal material derived by melting from the Monian rocks during subduction. Dewey (1969) was the first to suggest that the Monian orogeny resulted from the subduction of an oceanic plate towards the southeast (see fig. 20, diagram 1, after Thorpe, 1974). This simple model can of course account for the other late Precambrian events of England and Wales. Dewey interpreted the Monian and Longmyndian as correlative offshore and near-shore facies respectively, both formed on the south-eastern margin of a Precambrian Proto-Atlantic Ocean. Baker (1973) has reinterpreted them as the two sedimentary borders of a marginal Late Proterozoic ocean basin, (comparable with some in the modern West Pacific), pointing out that the source area of some of the Monian clastics lay to the northwest. Baker's interpretation is shown in Fig. 20, diagram 2. Anglesey (and Wexford) would have formed a micro-plate situated on the southern margin of "Iapetus". Thorpe enlarges further on the problem, bringing into the picture the late Proterozoic events in Shropshire and the Midlands. The Uriconian and Charnian calc-alkaline volcanics and the calc-alkaline plutonic complexes of the area could be explained by subduction of an oceanic plate either from an ocean northwest of Anglesey (Fig. 20.1) or from the southeast side of a margin basin south-east of Anglesey (Fig. 20.3 and Fig. 20.4). The absence of ophiolites and low temperature-high pressure phenomena in the Welsh Borderland is not necessarily a drawback as such preserved occurrences are fortunate anyway. One relic of ocean floor could however be represented in the Warren House volcanics, accreted into the south-east continental margin.

Recent work by Barber and Max (1979) has however shown that the Anglesey picture is more complex. They have resurrected the basal Gneisses and take the upper portions of the Monian (including the Gwna melanges) up into the Cambrian. Recent fossil finds in these melanges support this conclusion (see Muir et al, 1979).

Why did this early subduction occur? The leading edge of the southern continental plate (i.e. an edge facing France), freely moving southwards at first, must have encountered resistance. Such resistance could grow from spreading movements and subsequent orogeny in the mid-European ocean further south, connected perhaps ultimately with the Brioverian deposition and Cadomian deformation in N.W. France and Iberia.


## THE INFRACAMBRIAN ICE AGE

Glacial deposits (of Eocambrian or Infracambrian age) were first recognized in Northern Norway in 1891. They have been subsequently discovered in Spitzbergen, E.Greenland, Britain, C.Siberia, Australia, China, Africa, Brazil and Utah. The occurrences in Africa are readily explained by their high palaeolatitude position, but this explanation breaks down for areas like Scandinavia and Greenland. Palaeolatitude readings for rocks in Greenland give $8^{0}$ and those for Northern Norway give $4^{0}$. It is difficult to fit in glacial episodes (even if it is, as some suggest, a matter of drifting ice flows far removed from source) with equatorial latitudes. The picture is further complicated by the interbedded occurrences of dolomites suggesting rapid changes in temperature.

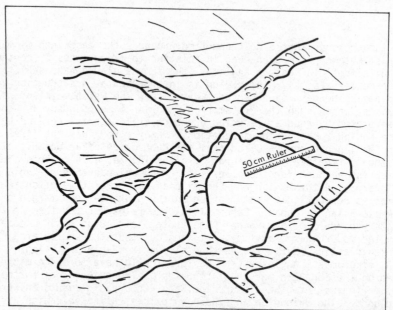

Fig. 22.  Polygonal sandstone wedges in the Garvellachs (based on Spencer, 1971, plate 8c).

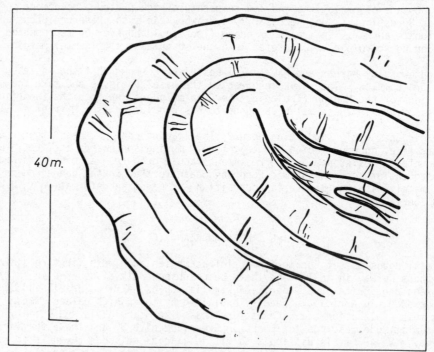

Fig. 23.  Sedimentary fragment 40m high in "the Great Breccia", one of the Garvellachs mixtites (after Spencer, 1971, plate 1).

These late Precambrian tillites in Scotland have been described by Spencer (1971). The tillite sequence can be traced over a distance of 700 km from N.E. Scotland to Connemara in West Ireland (fig. 21). The best known occurrence is at Port Askaig in Islay. Here this Middle Dalradian tillite sequence is 750 m thick. Abundant (and presumably far-travelled) granite stones, up to 1.5 m across, and large sedimentary fragments are contained in 47 "mixtites" (till-like beds with thicknesses from 50 cm to 65 m). The largest sedimentary fragment is 320 x 64 x 45 m. The mixtites are separated by siltstone, sandstone, conglomerate and dolomite interbeds (ranging from a few cms to 200 m in thickness). Certain individual mixtites can be correlated for a distance of 160 km between the Garvellach islands, Islay and Fanad. Sedimentary features of the interbeds include very variable palaeocurrents, beach conglomerates, wave-cut erosion surfaces, varves, outsize stones and drop-in structures produced by ice-rafting. Spencer has described polygonal sandstone wedges (fig. 22) of periglacial origin from the Garvellachs. The great size of sedimentary blocks in the "Great Breccia" mixtite of these Scottish islands (fig. 23) rules out a mudflow transportation (unless there was a very steep palaeoslope and for this there is absolutely no evidence). Spencer gives convincing reasons for rejecting non-glacial mechanisms such as turbidity currents and mudflows. Spencer recognises seventeen glacial advances and meltings in the Scottish tillite sequences.

Harland considers the Infracambrian glaciation to have been a world-wide phenomenon. One has to account for at least the presence of icebergs in equatorial waters. The widespread and fairly sudden appearance of fossils, together with a major marine transgression, in the Lower Cambrian, may have some considerable connection with a world-wide late Precambrian glaciation. The cause (or causes) of such a major glaciation presents further problems. Dineley (1974) puts forward an interesting possibility. He suggests that because of widespread and long-continued activity of blue-green algae in the late Precambrian seas, the proportion of carbon dioxide in the atmosphere dropped appreciably. The "greenhouse effect" disappeared when $CO_2$ was removed from the air and this reduced the infra-red retaining capacity of the atmosphere. Covers of snow and ice would then make matters worse by further reflecting the sun's energy.

The glaciation is perhaps a convenient marker in Britain for the boundary of Precambrian and Cambrian time and heralds the beginning of a geological period that was almost 100 million years in length. It is important to fully grasp the significance of this great length of time. In the first place it is long enough for changes of climate to have taken place and all the indications are that the British Cambrian was a time of warm conditions with even the southern margins of Britain no further south of the Equator than about 30° (fig. 24). With this length of time it makes it at least a little less difficult to explain the change in Scotland from glacial effects near the beginning of Cambrian times to the warm water depositional environment of the Durness carbonates of N.W. Scotland. Secondly, with this great length of time, the Proto-Atlantic Ocean could have widened appreciably during the Cambrian before starting to close near the end of the period. Fig. 24 suggests the picture for the Cambrian. Scotland and the north of Ireland could have been separated from the rest of Britain by an ocean several thousands of kilometers wide. On its immediate northern margin the Middle and Upper Daladian sediments were forming in a subsiding shelf-slope environment with increasing instability as U.Cambrian times approached when possibly temporary island arcs began to form oceanwards.

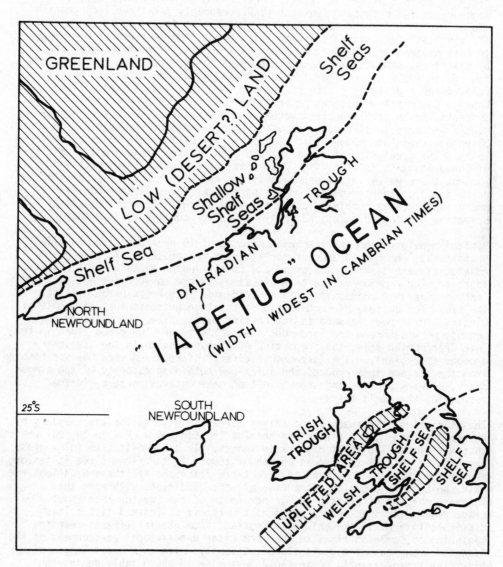

Fig. 24. "Iapetus" at end of Lower Cambrian times.

Traced landwards on this northern margin, the disturbed and often deeper
Dalradian waters gave way to very shallow shelf water covering what is today
the Hebrides and the northwestern fringe of the Scottish mainland.  The Cambrian waters had taken some time to flood on to this northwesternmost
margin, which was possibly still land when glacial ice-flows were dropping
erratics into the waters further south and east.  The first transgressions
to this "Hebridean" area deposited clean-washed sands, often riddled with
worm-burrows, but then came virtually a cessation of sediment supply (from
what perhaps was now a desert-type foreland).  Carbonate sheets began to
form very slowly in water rich in magnesium, as well as calcium, carbonate
and the Durness Limestone accumulated.  There were long cessations of even
this deposition - there might even have been gentle prolonged emergence
during Middle and Upper Cambrian times before renewed carbonate deposition
in the early Ordovician (the fossil evidence suggests the presence of only
Lower Cambrian and Lower Ordovician).

The southern margin of the ocean had a much more complex pattern of
deposition (fig. 24).  The end Precambrian subduction belt across Anglesey
and S.E. Ireland had given way to an isostatically-raised cordillera or
fault controlled horst formed of Monian (and underlying basement) rocks.
Continued instability along this S.Irish Sea region is reflected in this uplift and in the relatively rapid downwarping of the sea floor on either side.
As a result sediment eroded off the horst was deposited into the Welsh and
Irish troughs on either flank, forming the thick Harlech Dome and Bray successions respectively.  The Cambrian thickness in the Dome approaches
5000 m in places.  The base has been almost penetrated at last in a borehole
in the centre of the Dome where Arvonian volcanics were encountered beneath
the Dolwen Grits.  The thick Cambrian succession includes many turbidite
horizons (Rhinog Grits, Barmouth Grits, etc.).  The depth of the basin varied
appreciably at different times with relatively steep slopes causing rapid
flow.  The turbidity currents changed in direction with the ever-changing
submarine topography (Rushton, 1974).  At other times shallow water
deposition (perhaps following rapid infilling of troughs by turbidites)
occurred forming even partly enclosed basins (Manganese Shales).  Manganese
precipitation could have followed intensive weathering of spilitic or
keratophyric lavas (Monian or Arvonian) over the denuding Irish Sea Horst.
Further north,in Caernarvonshire, turbidite material was not as abundant, so
that after initial conglomerates and grits, muddy sediments (the famous
Bethesda-Llanberis slates) dominated the Lower Cambrian.  The Bronllwyd Grit
marks a coarser depositional beginning to Middle Cambrian times but was then
followed by uplift so that much of the middle division of the system is absent
(Wood, 1974).  Palaeoslopes (as indicated by slump folds) in this Caernarvonshire Cambrian sequence are towards the south, again suggesting the influence
of the Irish Sea Horst to the north and northwest.  The absence of Cambrian
in the North Pennine inliers could indicate a Lancashire-Yorkshire continuation of the same positive feature.  A source to the west or northwest is
indicated for the Pembrokeshire Cambrian sequence also.  Middle Cambrian is
well represented here but higher portions of the Lingula Flags are probably
missing (as is the Tremadocian) beneath an Ordovician overstep.  On the
northwest side of the Irish Sea Horst, thick Cambrian deposition again
occurred, especially in the Wicklow Hills where the Clara Group of the (older
named) Bray Series is almost certainly Cambrian and approaches 3000 m in
thickness.  Rocks beneath this group could also be of Cambrian age.  Faunas
are poor in this Irish Cambrian but the richness of trilobite faunas in the
Welsh Trough and their similarity to those in Scandinavia suggest free
migration from the Welsh area northeastwards.

The Welsh Trough, like the Dalradian one, gives way landwards to a shallow
shelf area, probably even wider on this southern side than on the northern
side of Iapetus.  The width is in fact difficult to assess in view of the
uncertainty about Cambrian (or Tremadoc) horizons in boreholes scattered over
England.  There may have been low-lying emerged areas at times over the S.E.
Midlands and S.E. England but at other times (especially Tremadocian) muddy
deposition, probably in shallow water, extended southeastwards well across
S.E. England.  In the Welsh Borders, Lower Cambrian clean-quartz sands soon
give way to glauconitic sands and these pale green sands could last until the
end of the Middle Cambrian in the Malverns.  In Shropshire however, they pass
up within the Lower Cambrian into a remarkable condensed carbonate sequence -
2 m of Comley Limestone - containing several distinct trilobite faunas
equating in time with very much thicker sequences in Scandinavia.  Slight
folding and erosion before Middle Cambrian deposition began in Shropshire
indicates some instability, perhaps because Shropshire lay (as it was to lie
also in Ordovician and Silurian times) on the hinge region beneath shelf and
trough.  This line, like that near N.W. Caernarvonshire, could reflect old
sutures in Precambrian basement, especially if a marginal basin opened and
closed between Shropshire and Anglesey in late Precambrian times.  Continued
instability in Shropshire is further indicated by the very incomplete and
thin character of the Upper Cambrian.  In Warwickshire, muds predominate
through Cambrian time, after 250 m of basal quartzites.  Some intermittent
breaks and phosphatic layers within the (Stockingford) muds suggest very
shallow water, yet conditions generally were sufficiently open-water to allow
free faunal communication with Scandinavia (and Wales) at times.

On the northwestern flanks of Iapetus it is possible to detect a land area
off the Hebrides, probably over Southern Greenland.  On the southern side of
the ocean, detection of land is not as easy.  The situation is made difficult
for a number of reasons.  Cambrian exposures are non-existent over southernmost England.  One does not know the palaeo-distance (great or small) between
southern England and N.France in Cambrian times.  Early and later Variscan
shortening and possible (though highly controversial) subduction of a pre-
Variscan ocean between Cornubia and Brittany add to the problems of reconstructing the "southern land area" for Iapetus in Cambrian times.  It seems
however reasonable to postulate land between Cornubia and N.W. France (see
Sutton and Watson, 1970 and Renouf, 1974, fig. 3C).  A land area
("Domnonaea") lay to the west of the Channel Isles.  In Ordovician times,
this West Channel land mass was to be gradually attacked from the south by a
transgressing sea (Renouf, 1974, fig. 3D).


                         THE  GATHERING  STORM

Years ago it was traditional to think of the Caledonian Earth Movements as
having taken place in a brief interval between the end of the Silurian and
the laying down of the Old Red Sandstone.  This was a long-standing belief
based on the unconformable relationship of the Old Red Sandstone to various
members of the Lower Palaeozoic or even the Precambrian.  Nowadays it is
believed that earth movements are protracted affairs and in fact the
Caledonian Orogenic Cycle lasted several millions of years - in fact the
length of time represented by the opening and closing of Iapetus.  Main
climaxes of the Caledonian Orogeny in "Caledonia", i.e. Scotland, occurred
in fact in early Ordovician times.  This is the time when the Moine and the
Dalradian were intensely folded and metamorphosed to make the first

## A. <u>EARLY ORDOVICIAN</u>

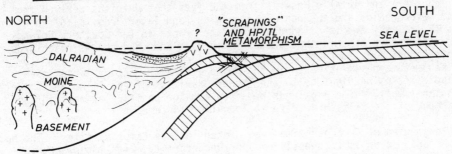

## B. <u>LATE ORDOVICIAN</u>

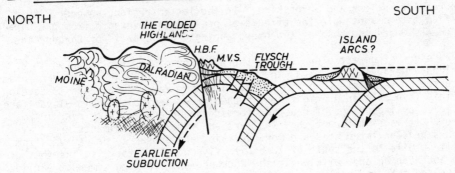

## C. <u>MID-SILURIAN</u>

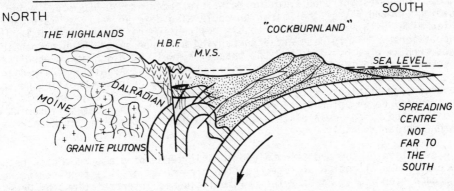

Fig. 25.   Evolution of Scotland in Lower Palaeozoic times
(partly after Mitchell and McKerrow, 1974). H.B.F. Highland
Boundary Fault;  M.V.S. Midland Valley of Scotland.  Dotted
shading - flysch; Striped shading - ocean crust.

Caledonian Highlands.  We know this from radiometric datings (see earlier in
this chapter) and we also know it from the fact that detritus derived from
these metamorphics are abundant in Arenigian (and later Ordovician) rocks in
Mayo (W.Ireland) and in Caradocian and later rocks in the Southern Uplands.
It can therefore be said that the Scottish Highlands had become "Highlands"
by mid-Ordovician times and although given further "shots in the arm" (by
further metamorphism at depth in the Moine at least - and by large scale
periodic intrusion right up to the "Newer Granites" of end Silurian to early
Devonian times), these Highlands remained a major positive element in the
geography of the British area for a long time to come.

The long quiet of Cambrian times was broken in late Cambrian times.  The base
of the Ordovician is frequently a major unconformity, with important overstep, in many areas.  In North Wales, as George has shown, the Ordovician
rapidly oversteps northwestwards from the Harlech Dome, across St. Tudwal's
to N.W.Caernarvonshire and Anglesey where, in the main, the Ordovician rests
on Monian.  In Pembrokeshire also there is an important break, and again in
the Pennine inliers and Shropshire.  In the latter region, however, it is
difficult to disentangle the effects of pre-Arenig movements from pre-Caradoc
movements where the base of the overstepping Ordovician is of Caradoc age.

Early signs of unrest along the northern border of Iapetus occurred in W.
Ireland.  Here, in Connemara and Mayo, "Grampian" folding and metamorphism
occurred in very late Cambrian times.  Then the Mayo Basin began to develop
in this fold belt in Arenig times.  Ophiolites, including pillow-lavas, are
followed by inpourings of clastics forming an Ordovician pile over 10 km
thick.  It is significant that the earlier Ordovician sediments are greywackes and the later ones are more deltaic, and appear to be derived from a
built-up pile to the south of turbidite flysch.  The story is like that of
the Southern Uplands with an Irish "Cockburnland" forming somewhat earlier in
W.Ireland than in Scotland.  Late Cambrian to very early Ordovician subduction
must have occurred along this Irish sector of North Iapetus.

In the Scottish Highlands, deep processes were growing even during Cambrian
times.  Granites intruded into the Moine are dated at 540-550 m.y.  Major
folding ($f_1$ and $f_2$) of the Dalradian could also date as early as late Cambrian times with major nappes probably delineating different later Dalradian
successions.  This folding and accompanying high pressure - high temperature
metamorphism grew outwards and the phases may well be diachronous when traced
southwards to the Highland Border region where almost Arenigian "Dalradian"
are involved in nappe folding.  Along the Highland Boundary Fault occur carbonated serpentinites, with cherts and jasperised pillow lavas and this wedge
must surely be an upthrust of "obducted" wedge of ocean crust splintering off
a major subducting ocean plate disappearing beneath the Grampian Highlands in
late Cambrian to early Ordovician times.

The evolution of the Scottish Caledonides in terms of plate tectonics has
received much attention of late, but two treatments of this fascinating new
approach stand out.  Firstly the original classic interpretations by Dewey
(1969) and by Dewey and Pankhurst (1970).  (Dewey's interpretation has been
further attractively presented in the Open University course "Historical
Geology").  Secondly, a clear account of the evolution of the Caledonian belt
is given by Mitchell and McKerrow (1975).  At the same time they draw attention to the comparative evolution and pattern of this ancient Scottish belt
of orogeny and the present day structural pattern and evolution of Burma.  An
eastward pattern of belts from the Bay of Bengal to the E.Highlands of Burma

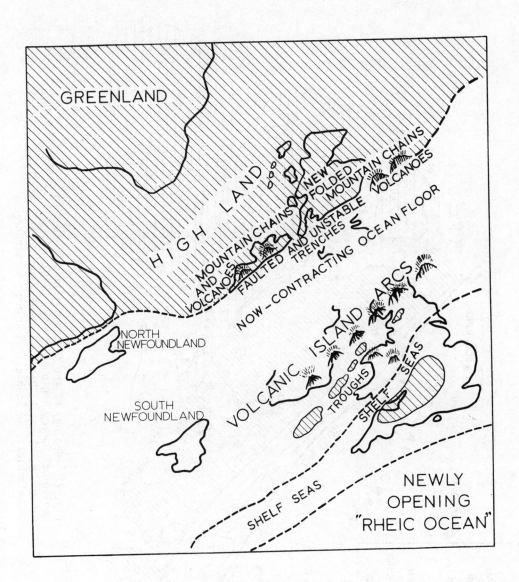

Fig. 26.  "Iapetus" in Lower Ordovician times.

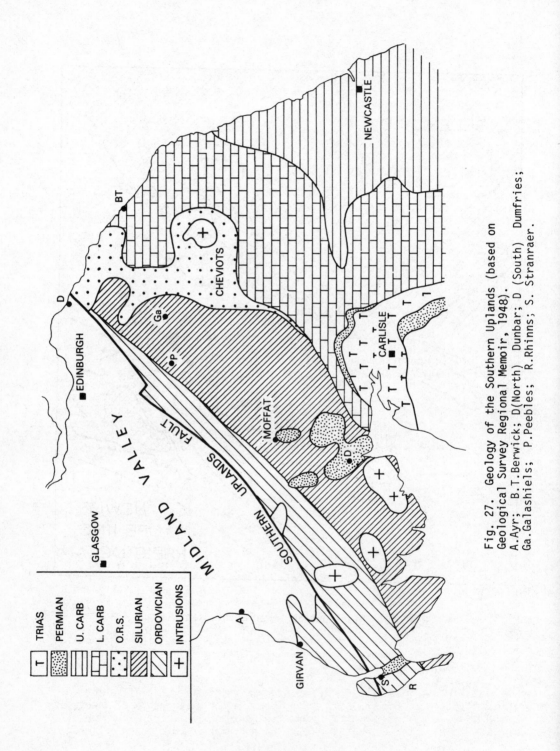

Fig. 27. Geology of the Southern Uplands (based on Geological Survey Regional Memoir, 1948). A.Ayr; B.T.Berwick; D(North) Dunbar; D (South) Dumfries; Ga.Galashiels; P.Peebles; R.Rhinns; S. Stranraer.

is compared with the northward change in the Scottish Caledonian Orogen from the Southern Uplands to the Midland Valley and the Grampians. Fig.25 is in part adapted from Mitchell and McKerrow's interpretation and attempts to show the stages in the evolution of the Grampians, Midland Valley and Southern Uplands areas from late Cambrian to mid-Silurian times. By late Cambrian times, the northern moving ocean plate of Iapetus was now meeting resistance and was beginning to subduct beneath the Grampian portion of the northern continental plate (fig. 25A). Gabbros intruded into Dalradian and dated at about 510 m.y. suggest that some deeper melting process had already begun. By earliest Arenig times, the subduction was resulting in widespread compression and heating of the Dalradian (over Moine) pile and continuing resistance was buckling the sediments into more and more complex patterns with over-folding and/or thrusting into two outward directions (southwards near the site of the Highland Border). Obducted slices preserved (at high structural levels) portions of ocean crust and overlying oceanic sediments, with high pressure-low temperature metamorphism resulting in serpentinites being caught up near the line of the present Highland Boundary Fault (e.g. near Loch Lomond and Stonehaven). Granite emplacement into the uplifting Grampian pile continued into lower to middle Ordovician times. During Arenig to Llanvirn times, spilitic lavas with shales and cherts were deposited near the Highland Border and probably also over vast areas to the south (Girvan and Moffat and the other Southern Uplands inliers). This area was however probably in a greater state of unrest than the simple correlation of rock types over a vast area would have one believe. Eclogites, serpentinites, gabbros and blue schist metamorphics occur near Girvan with complete faulted and thrust field patterns. They appear to be of Arenig (perhaps pre-Middle Arenig) age and must represent upthrust (obducted?) splinters. This obduction could of course have been a southward remnant of that along the Highland Border. It is more likely however to represent a separate obduction suggesting newer subduction lines in the vicinity of the present day Southern Uplands Fault. This raises the possibility (fig. 25B) that only ocean crust under-lies the Arenigian of at least the southern side of the Midland Valley.

Llandeilian deposits may be absent in the Girvan area or may be very thin. If they are thin, then this attenuation might represent the beginnings of a subduction trench in that area in mid-Ordovician times (such trenches in front of island-arcs or mountain fronts often have only a thin film of sediment, at least at first). With the growing new subduction near the northern edge of the Southern Uplands, calc-alkaline magmas began to rise through the cover to the north, probably forming andesite piles somewhere in the northern portion of the Midland Valley (fig. 25B). Andesitic detritus (plus Dalradian and other metamorphics) occur widely in the Bala turbidites of the Southern Uplands (including Girvan). By Caradocian times, great lateral changes were beginning to occur in the Girvan area. Shelf con-glomerates and limestones in the north of the Girvan area pass southwards rapidly into a 5000 m sequence of turbidites and shown by Williams (1962) to have accumulated along a fault affected zone with southeastward down-throwing contemporaneous fractures (fig. 25B). Further to the south, a second subduction belt may have been developing near the site of deposition of the Wrae (Bala) Limestone with andesites and rhyolites forming an island arc (fig. 25B).

Fig. 26 attempts to represent in one Ordovician picture what is really a story of continued change, but, to sum up, the northern margin of Iapetus in the Ordovician was now an area of great unrest as the northern continental block resisted with increasing vigour the northward moving oceanic plate.

Subduction belt followed subduction belt with faulted shelf margins and infilled trenches. It seems inevitable that the spreading centre was moving
asymmetrically towards the northern side of Iapetus by now.

What now of the southern side of the ocean? In the Lake District and the
Isle of Man great thicknesses of turbidites were deposited on the northern
side of the old Irish Sea Horst as the Manx and Skiddaw "Slates" (muds were
in fact subordinate to coarser greywackes) in lower Ordovician times. Thickness estimates range from 2500 to 6000 metres and there could be similar
Cambrian lithologies beneath. Still further south in Wales, Arenig and
Llanvirn times saw variable coarse deposition at first giving way to finer
muds (with graptolite faunas) but with widespread vulcanicity in several
areas especially Cader Idris, Arenig Mountain, the Arans, South Snowdonia,
Trefgarn (N.Pembrokeshire), Strumble (Fishguard) and the Builth area.
Rhyolite and andesite lavas and tuffs are particularly common whilst pillow
lavas are very well represented in the Fishguard (Strumble Head) area. It
is difficult to get away from the notion of an island-arc environment with
these calc-alkaline magmas dominating the Welsh basin of deposition in Lower
Ordovician times. It is difficult also to resist the temptation of resur-
recting a reactivated Anglesey-S.E. Ireland subduction belt to account for
this Welsh igneous activity. Further to the east, the lower to middle
Ordovician sediments of the Shelve area of Shropshire, though graptolitic in
some horizons, are really a shelf facies and the volcanic outbursts are more
sporadic and thinner.

The ocean was therefore by mid-Ordovician times meeting resistance on both
margins and was now contracting appreciably. That contraction was accentuated
before Caradocian times by the development of another subduction zone (dipping southwards) in the vicinity of the Lake District (and across to the
Dublin area). The age of the calc-alkaline pile that developed above this
subducting belt is problematical but is younger than a bifidus position and
older than Caradoc. At least 3000 to 4000 metres of volcanic lavas and tuffs
accumulated in this Borrowdale Volcanic Series with andesitic types predominating (but rhyolites are important too). The island arc pile in the
Lake District must have built up appreciably and was to suffer folding and
erosion before Upper Ordovician deposition was renewed. Folding and faulting
was mainly along NNE-SSW lines, the folded volcanics being then transgressed
by thin U.Ordovician mudstones, impure limestones, ashes and rhyolite flows.
The thin quiet character of these deposits in the Lake District contrasts
however with the thicker Bala deposition in North Wales and in S.E. Ireland -
areas where intense volcanic activity was to break out again. The thick
Snowdon Volcanic Series (largely ignimbritic) is of Caradocian age. In
E.Ireland submarine vulcanicity broke out north and west of Dublin and
between Wicklow and the Waterford area with rhyolites becoming important in
that southern area.

It is difficult to account for the return to relatively quiet, passive conditions in the Lake District in U.Ordovician times whilst renewed igneous vigour
returned to areas further south. One clue may be the pre-Caradocian folding
and (faulting) which affected a much wider area of England and Wales
(especially the Welsh Border) than was hitherto suspected. It seems almost as
if the Lake District subduction drive met some kind of temporary barrier and
that after some brief halting and shuddering it carried on subducting down
ocean plate with melting now rising mainly in the North Wales-S.E. Ireland
segment of crust. Another alternative is that the Lake District subduction
belt stopped acting for some reason in U.Ordovician times but resistance way
to the south reactivated once more the earlier Anglesey-S.E. Ireland zone.

The Welsh trough gave way southeastwards (rather suddenly at about a line
from Berwyn to Shelve to Builth) to shelf conditions though pre-Caradocian
uplift and erosion limited E.Shropshire  preserved sediments to the Caradocian. That Ordovician deposition did extend well into the Midlands and
beyond however has been shown by some critical boreholes more recently.
Nevertheless land areas did persist somewhere to the south.  No Ordovician
has been proved over S.E. Wales or Southern England.  New transgressions
from the other side however, i.e. from Brittany and Normandy were by mid-
Ordovician times now extending northwards to an area of rocks now represented
by localities in South Cornwall (these may then have been much further south
of their present English position).  The Gorran Quartzites exposed at Ger-
rans Bay and Veryans Bay can now be correlated (Sadler, 1974) on the basis of
trilobites with the Llandeilian Gres de May inferieur of Normandy and with
rocks in Finistere.  The Cornish faunas contrast strongly with Anglo-Welsh
areas but compare with those of Armorica, Iberia and North Africa.

## THE OCEAN CLOSES

And so to Silurian times, a period that was to see the final rapid closings
of the Proto-Atlantic Ocean.  The Girvan turbidite flood soon began to reach
other Southern Uplands areas such as Moffat (40 m of Bala muds gave way to
over 6 km of greywackes in a short period in the early Silurian).  Some of
this material could have been derived from a southerly positioned rise.  With
continued subduction under the Southern Uplands and Midland Valley, great
northward dipping thrust slices were being built up which eventually, by
Wenlock times, formed an important rise and source of debris to which Walton
has given the name "Cockburnland" (fig. 25C and fig. 28).  Over this region
progressively younger turbidites occur southwards in each tectonic unit.
From Cockburnland, debris spread northwards into the mid-Silurian trough of
the Midland Valley and southwards into more southern areas of the Southern
Uplands.  Andesites were probably still being erupted in the north or north-
eastern parts of the Midland Valley as andesitic detritus is abundant in
Scottish turbidites throughout lower and middle Silurian times.  The considerable thickness of continental crust under the Southern Uplands can, according
to Mitchell and McKerrow, be explained either by the huge turbidite piles
stacked on one another and metamorphosed at depth or by accumulations over
the Lake District island arc being eventually (in the final ocean closures
and tight suturings) being underthrust beneath the Southern Uplands).

On the other side of the now closing ocean, reactivation of the Irish Sea
zone pushed up new supply areas which furnished abundant debris for the still
subsiding Welsh trough.  Turbidites flooded into Cardiganshire in lower
Silurian times (Wood and Smith have shown the direction of transport of the
thick Aberystwyth Grits to be north-north-eastwards) and into Denbighshire
and Montgomeryshire in later Silurian times (from westward and more southward
sources respectively).  In mid-Pembrokeshire, at the close of Ordovician and
during the earlier part of Silurian times, the thick sequence of the Skomer
(and Marloes) Volcanic Series accumulated, in places up to 1000 m thick.
The lavas range from basic to acid but there are indications of a partly subaerial environment with reddening of flow surfaces at Marloes.  This important
volcanic outburst (once thought to have taken place in the early Ordovician)
must again represent some off-Pembrokeshire subduction activity.   Similar
activity (not necessarily along the same zone) has to be postulated in much
later Silurian times to account for volcanic outbursts in the Dingle Peninsula
of S.W. Ireland.  To the east and south of the Welsh Trough, shelf conditions

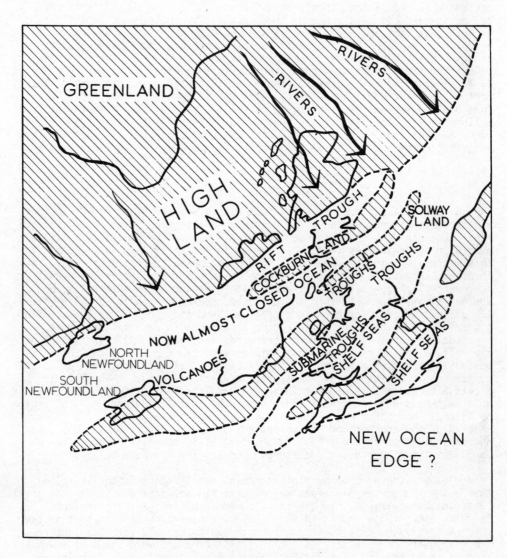

Fig. 28.  "Iapetus" in Middle Silurian times.

prevailed in South Pembrokeshire, S.E. Wales, the Welsh Borderland, Central England and the Bristol area. In Wenlock and Ludlow times, carbonate deposits accumulated, with reefs, in the classic areas bearing the same names. The shelf-slope boundary in the Welsh Borderland was now relatively steep and active, as shown by the slipped masses of Aymestry Limestone caught up in graptolitic muds in the Leintwardine area. Variable low emergent land areas existed at first in the Silurian over eastern and southern England but by Wenlock and later times, a Midland-Wiltshire barrier appears to have separated two shelf regions with the eastern shelf passing Kent-wise into deeper muddy waters (boreholes have recorded this information).

Towards the close of Silurian times, final closures of the ocean, and its marginal troughs and slopes, were beginning to compress the Lower Palaeozoic sediments and volcanics of the Welsh and Irish areas. In the Welsh Borderland shelf area, this compression seems to have occurred firstly in early Silurian times and again in the mid-Devonian, with absolute conformity between Ludlow and Downtonian sediments. In the Lake District thick Ludlovian infilling (about 4000m) of that trough region continued until very late in Silurian times and the folding must be virtually all of L.Devonian age. The Wenlock and Ludlow turbidites of the Austwick-Horton inliers show very variable northerly or southerly source directions with time.

## Suggested Further Reading

Allen, P., Sutton, J. & Watson, J.V. 1974. Torridonian tourmaline-quartz pebbles and the Precambrian crust northwest of Britain. Jl. geol. Soc. Lond. 130, 85.

Barber, A.J. & Max, M.D. 1979. A new look at the Mona Complex (Anglesey, North Wales). Jl. geol. Soc. Lond. 136, 407.

Bath, A.H. 1974. New isotopic age data on rocks from the Long Mynd, Shropshire, Jl. geol. Soc. Lond. 130,567.

Dewey, J.F. 1969. The evolution of the Appalachian-Caledonian orogen. Nature, Lond. 222, 124.

Dewey, J.F. 1971. A model for the Lower Palaeozoic evolution of the southern margin of the early Caledonides of Scotland and Ireland. Scott. J. Geol. 7, 219.

Dewey, J.F. & Pankhurst, R.J. 1970. The evolution of the Scottish Caledonides in relation to their isotopic age pattern. Trans. R. Soc. Edinb. 68, 361.

Gass, I.G. et al. 1972. Historical Geology. S23, Block 6. Open Univ. Press. Bletchley.

Holland, C.H. 1974. Lower Palaeozoic rocks of the World, Vol. 2 Cambrian of the British Isles, Norden and Spitzbergen. London.

Hughes, N.F. 1973. Organisms and Continents through Time. Special Papers in Palaeontology, No. 12. Pal. Ass. London.

Kelling, G. 1962. The Petrology and sedimentation of Upper Ordovician rocks in the Rhinns of Galloway, South-West Scotland. Trans. R. Soc. Edinb. 65, 107.

McKerrow, W.S. & Zeigler, A.M. 1974. Silurian Palaeogeographic Development of the Proto-Atlantic Ocean.

Mitchell, A.H.G. & McKerrow, W.S. 1975. Analogous Evolution of the Burma Orogen and the Scottish Caledonides. Bull. geol. Soc. Am. 86/3, 305.

Muir, M.D. and others, 1979. Palaeontological evidence for the age of some supposedly Precambrian rocks in Anglesey, North Wales. Jl. geol. Soc. Lond. 136, 61.

Read, H.H. & Watson, J. 1975.  Introduction to Geology, Vol. 2. Earth History
    Part II.  Macmillan Press.  London and Basingstoke.
Renouf, J.T. 1974.  The Proterozoic and Palaeozoic Development of the Armorican
    and Cornubian Provinces.  Proc. Ussher Soc. 3, 6.
Spencer, A.M. 1971.  Late Precambrian Glaciation in Scotland.  Mem. Geol.
    Soc. Lond. No. 6.
Tarling, D.H. 1974.  A palaeomagnetic study of Eocambrian tillites in Scotland.
    Jl. geol. Soc. Lond. 130, 163.
Thorpe, R.S. 1974.  Aspects of magmatism and plate tectonics in the Precambrian
    of England and Wales.  Geol. Journ. 9, 115.
Williams, G.E. 1969.  Petrography and origin of pebbles from Torridonian
    strata (late Precambrian), northwest Scotland.  In: North Atlantic - Geology
    and Continental Drift.  Mem. 12, Amer. Assoc. petrol. Geol. 609.
Wood, D.S. 1974.  Ophiolites, melanges, blueschists and igimbrites: Early
    Caledonian subduction in Wales.  In: Modern and Ancient Geosynclinal
    Sedimentation.  (Ed. Dott and Shaver). Tulsa.
Ziegler, A.M. & McKerrow, W.S. 1975.  Silurian Marine Red Beds.
    Amer. J. Sc. 275, 31.

## CHAPTER 3

# "Even Further Back... ? "

## THE LEWISIAN COMPLEX

In the previous chapter the story of the British area was taken back to about 1000 million years ago. To venture even further back into geological time with the British story it is necessary to go to the extreme northwest of Scotland, to examine rocks which in some cases may be over 3000 million years old. It is extremely fortunate that a small remnant of this very ancient floor is still preserved on the modern surface of Britain. Though small in extent (see fig. 29.), the remnant has nevertheless revealed a long and extremely complex history. The rocks are very much altered, having been affected by severe folding and by great changes of heat and pressure. In many cases it is difficult to trace the original nature of the rocks. Methods and techniques used in interpreting the nature and order of formation of younger geological units cannot be applied to these old rocks; in fact chemical methods often play an important role. Radiometric dating also makes an important contribution, though the interpretation of the various radiometric readings can itself be a difficult task, presenting additional problems. These radiometric dates can be influenced by so many environmental factors - intrusive activity, various phases of earth movement and metamorphism, etc.

Lewisian rocks are the oldest in the British Isles. They form an irregular, mainly coastal, fringe, up to 40 Km wide, along the northwest corner of the Scottish mainland from Cape Wrath to Applecross. They also form the whole of the Outer Hebrides, with the exception of small areas of New Red Sandstone on the east side of Lewis. Further Lewisian remnants form the islands of Tiree Coll and Iona and parts of Raasay and Islay. Inistrahull, an island off the north coast of Donegal, Ireland, is a further Lewisian relic. A number of Lewisian inliers (overlain by the Moine Complex) occur on the Scottish mainland to the east of the Moine Thrust. The largest of these areas is the Glenelg inlier with an extension into the Sleat peninsula of Skye.

The Lewisian is a metamorphic assemblage composed of two principal units, an older Scourian Complex and a younger Laxfordian Complex. This division was first established (in 1951) by Sutton and Watson and was confirmed by the first radiometric datings published between 1959 and 1964. Each complex represents an orogenic belt with a very long succession of events. The Scourian covers a period from about 2900 million to 2200 million years ago, the Laxfordian from 2200 million to about 1500 million years ago. A convenient separation of the two units was the intrusion of a great swarm of basic dykes, trending NW-SE to WNW-ESE. These occur particularly in the southern parts of the Outer Hebrides and in the central portion of the mainland fringe. The Scourian and Laxfordian complexes are each best described as orogenic cycles. They each represent much longer times however than the various Phanerozoic orogenic cycles and perhaps the more major crustal term "chelogenic cycle" is better applied to the Scourian or Laxfordian units. Each of these complexes represents after all some 700 million years of time - longer than from the beginning of the Cambrian to the present day.

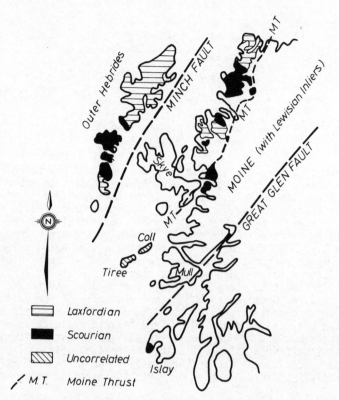

Fig. 29.   The Lewisian of N.W. Scotland (after Dalziel, 1969).

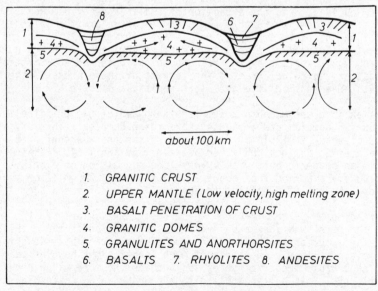

1.   GRANITIC CRUST
2.   UPPER MANTLE (Low velocity, high melting zone)
3.   BASALT PENETRATION OF CRUST
4.   GRANITIC DOMES
5.   GRANULITES AND ANORTHORSITES
6.   BASALTS   7. RHYOLITES   8. ANDESITES

Fig. 30.   Early crustal processes (after Fyfe, 1974).

The Scourian and Laxfordian are formed largely of grey banded gneisses,
rocks which have been highly metamorphosed and often mignatised at high temperatures and pressures.  It is difficult in fact to make out their previous
origin.  Metamorphosed sediments are represented by mica-schists, calc-
silicate rocks and marble.  They occur particularly in the Gairloch and Loch
Maree districts and in South Harris (Outer Hebrides).  Various igneous masses
- granites and pegmatites - are found in the Laxfordian complex.

The main Scourian mass of the Scottish mainland occurs along the Scourie-
Lochinver fringe.  Here the highly altered gneisses contain no obvious relics
of metasediments.  Grey banded gneisses of acid-intermediate composition form
the bulk of the area.  They are largely characterised by the occurrence of
the mineral hypersthene.  Metamorphism took place in the granulite facies
under very high temperatures and pressures.  There are certain comparisons
with the charnockite facies seen in very deeply-eroded orogenic belts
throughout the world.  Pegmatite veins are rare indicating a "dry" environment for the metamorphism.  The Scourian gneisses enclose a number of more
basic bodies.  Some are pure pyroxene or olivine rocks.  Banding could
represent original layering in what could have been ultrabasic igneous bodies.
Other basic masses, usually larger zones, bands or lenses are pyroxene-
granulites, often full of red garnets.  The structural trend in this "central
belt" of the mainland Lewisian is variable but can be frequently between
NE-SW and E-W.  This is in marked contrast to the predominant NW-SE trend of
the Laxfordian areas on this mainland fringe.

Smaller Scourian areas occur further south on this fringe, as for example on
the north side of Loch Torridon.  Kyanite gneisses in the Carnmore area are
also probably part of the Scourian assemblage.  The Scourian dating is best
afforded by the way in which the trend of these old gneisses is cut by
little-altered basic dykes.  Calcareous granulites here could be migmatised
metasediments.  It is however in the Outer Hebrides that the other main area
of Scourian complex is to be found.  Here on South Uist, North Uist and South
Harris, the structural pattern pre-dates basic dykes and sills considered by
Dearnley to be of the same age as the post-Scourian dyke swarm of the mainland.  However the areas have suffered further Laxfordian metamorphism on two
occasions and only relics of the initial granulite facies occur.  On South
Harris, metamorphosed tonalites and anorthosites have been assigned to the
Scourian or to the dyke swarm episode, even though they bear Laxfordian
effects.  Metasediments ( pelites , marble, calcareous and quartzitic bands)
on South Harris which are invaded by these altered igneous masses could
therefore be either pre-Scourian or Scourian in age.  Migmatites in the
Glenelg inlier of the mainland are again cut by dyke-like amphibolite bands
and could also be Scourian.  A similar relationship could apply in Islay and
rocks resembling the Scourian gneisses of Glenelg occur on Iona, Tiree and
Coll.

The post-Scourian dyke-swarm comprises tholeiitic dolerites and ultrabasics.
A consistent NW-SE or WNW-ESE trend characterises the swarm on the mainland.
The N-S extent of the area affected by the dykes is nearly 300 Km.  They can
occur at at least three or four per kilometre.  The dykes must represent a
major regional intrusive event rather than some local igneous focal point.
This broader implication is supported by the occurrence of contemporaneous
swarm activity in Canada and Greenland.   The dykes have yielded ages of
2200 million years in the Lochinver area but over the whole swarm area the
dyke activity could cover a longer time span.  Some dyke activity could have
even extended into early phases of Laxfordian orogenesis.  Dearnley regarded

the South Harris igneous complex as a layered plutonic representative of this
late to post-Scourian dyke swarm. The juxtaposition of this plutonic body
to the South Harris metasediments lends further support to the possible pre-
Scourian age of those sediments. They could be the oldest part of Britain.

The Laxfordian Complex predominates in the extreme NW corner of the Scottish
mainland and in the large island of Lewis. Other areas occur in the southern
portion of the mainland fringe. Over much of the Laxfordian areas, the
gneisses were already in a gneissose condition before the advent of Laxfordian orogeny and metamorphism. Laxfordian rocks have therefore in many
cases suffered Scourian and Laxfordian metamorphism. These abundantly-
occurring polycyclic metamorphics are commonly of amphibolite facies. Many
are invaded by pegmatites and granite gneiss. The Laxfordian areas are also
however characterised by remnants which escaped the bulk of Laxfordian
orogenesis and metamorphism (similar in other words to the main central
Scourian tract of the mainland). They often form upfolded structures with
more affected Laxfordian downwarps in between. In the more affected areas
the post-Scourian dykes have been distorted and changed to amphibolites or
granulites. One sees comparisons of this unaffected Scourian within
affected Laxfordian with the various ancient massifs (for example, Mont
Blanc) caught up within the great Tertiary nappe areas of the Alpine chains.
The largest Scourian massif would have been the central tract of the mainland
fringe (an even larger mass if the Outer Hebrides Scourian area is its displaced continuation, as suggested by Dearnley in 1962). The northern edge of
the mainland Scourian massif is marked by north-westerly tending shear belts
giving way to a predominant north-westerly striking foliation in the gneisses
which now include numerous pegmatite veins and granitic sheets.

In the last ten to fifteen years a considerable amount of work has been
carried out on the Lewisian of the mainland and the Outer Hebrides. It would
not be surprising that in an overall region of rocks of such complexity and
with such a long and diverse history, detailed individual studies of smaller
areas within this Lewisian tract would reveal discrepancies in metamorphic
and structural evolution and a wide variety of radiometric dates. Some mainland areas have revealed an intermediate orogenesis (the Inverian) affecting
the Scourian. This intermediate phase was followed very closely by the dyke
intrusion and to some extent might even have overlapped. One difficulty here
is to decide whether these Inverian movements were in fact very late Scourian
or very early Laxfordian deformations. Again they did not affect all areas
of the Scottish region. The increasing number of radiometric dates available
have shown how prolonged were the Scourian and Laxfordian cycles. The early
radiometric figures falling into two distinct sets (2500-2200 m.y. from
Scourian pegmatites and 1650-1200 m.y. from Laxfordian gneisses) now present
too simple a picture and moreover minimise the ages of the main phases of
both Scourian (2750 m.y.) and Laxfordian (1975 m.y.) metamorphism. In order
to clarify the position and also in order to show the varying detailed
evolutions of different areas, two recent mainland studies will be referred
to: (a) the Lochinver area (type area for the Inverian metamorphism) and
(b) the Gairloch and Loch Maree areas, with their interesting metasediment
assemblages.

(a)  The Lochinver district has been recently described (1974) by Evans and
     St. Lambert who deduced the following chronology:-

c.2900 m.y.  primary age of pyroxene granulites.

> 2600 m.y.  termination of pyroxene granulite facies metamorphism,
             formation of the pyroxene gneisses of the Scourie
             assemblage.

2540 m.y.    intrusion of potash pegmatites in both the mainland
             Lewisian and the Outer Hebrides.

2310 m.y.    further potash pegmatites or recrystallisation of first set.

2310-2200 m.y. amphibolite facies metamorphism affecting WNW striking
             vertical belts up to 2 Km wide;  development of isoclinal
             folds, new gneiss banding and destruction of all pre-
             existing structures.  Formation of the Inver assemblage by
             hydration of the Scourie assemblage.

2200 m.y.    intrusion of ultramafic and mafic dykes into hot country
             rock followed closely by their autometamorphism.  The last
             members of the suite retain pyroxene mineralogy.

c.1850 m.y.  onset of decline of amphibolite facies metamorphism of the
             Laxford assemblage followed by slow cooling and foliated
             granite and intrusive pegmatite activity.   Brittle
             deformation of pyroxene granulites and amphibolites.

c.1600 m.y.  cessation of pegmatite activity, closure of biotite and
             hornblende to argon loss.  Termination of brittle shear
             activity.

c.1400 m.y.  closure of biotite to Rb and, or, Sr migration.


(b)  Rb-Sr isotopic studies of the metasediments and gneisses in the
     Gairloch - Loch Maree - Carnmore areas by Bikerman, Bowes and Van
     Breeman were published in 1975.  These authors put forward the follow-
     ing chronological outline for that region:

> 2800 m.y.  deposition of sediments and/or lavas.

2800-2700 m.y. dynamothermal metamorphism and tectonics (Scourian);
             formation of gneisses.

After 2200 m.y. deposition of Loch Maree Group and Gairloch sediments.

1975 m.y.    dynamothermal metamorphism and tectonics (main Laxfordian
             phase);  formation of schists and gneisses;
             Then thrusting and folding with formation of synforms and
             antiforms.

1750 m.y.    granite and pegmatite replacement; reheating;
             Then open folding.

1500 m.y.    epeirogenic uplift and cooling.

Bikerman, Bowes and Van Breeman envisage a Laxfordian orogenic cycle consisting of a depositional episode 2.2-2.0 b.y. ago, an orogenic episode
2.0-1.7 b.y. ago and an epeirogenic episode 1.7-1.5 b.y. ago.

There are still however many problems to solve concerning the Lewisian.
Earlier Scourian events are not easy to disentangle and the original nature
of the Scourian is a matter of controversy.  Dearnley and Dunning in 1968
concluded that much of the original material of the grey gneissic complex in
the Outer Hebrides may represent the products of a pre-Scourian orogenic
cycle of possible Katarchaean age (over 3000 million years).  They found no
evidence for igneous origins for the acid parts of the old grey gneiss.
Banding suggested a supracrustal series (of greywacke or leptite type) with
layers of basic volcanics.  Watson and Lisle have identified (1973) four
major assemblages in the pre-Laxfordian complex of the Outer Hebrides.
Assemblage 4 is the Scourie dyke suite.  Assemblage 3 is a late Scourian
intrusive suite emplaced in Assemblages 1 and 2 but themselves intruded
by the dyke suite.  These third assemblage intrusives range from ultramafic
and mafic bodies, rich in pyroxene or hornblende, to foliated granites and
potash-felspar pegmatites.  In the island of Barra, where Laxfordian metamorphism was slight, they have yielded isotopic ages ranging from 2450-2600
m.y.  Assemblage 1 is composed mainly of quartz felspathic gneisses (the grey
gneisses) with only a small percentage formed by large bodies of basic
material within the gneiss complex.  Watson and Lisle have estimated that
this first assemblage probably formed 80% of the Outer Hebridean Lewisian
prior to the Laxfordian regenerative episodes.  The second assemblage comprises metamorphosed supracrystal rocks with associated gneisses.  These
supracrustals include quartizitic, semi-pelitic, calcareous and graphitic
gneisses with occasional marble.  These metasediments are associated with
metamorphised igneous rocks (amphibolites, pyroxene-granulites, serpentinites,
hornblende-schists).  Watson and Lisle point out the similarity of what is a
supracrustal pile incorporating basic volcanics plus high-level basic
intrusives with the famous "greenstone-belts" of the great Archaean belts of
the World.  The dominance of basics contrasts with the acid-intermediate
character of the first assemblage.  The relationships of the first and second
assemblages are difficult to unravel.  The gneisses of assemblage 1 could be
migmatised derivatives of supracrustals like those of assemblage 2.  On the
other hand, the supracrustals could have accumulated on a gneissose or
granitic basement now represented by the first assemblage.  Because portions
of the second assemblage are themselves now gneissose the unconformable
junction between cover and basement would be obliterated and would probably
lie within the first assemblage.  These are important questions.  These first
two assemblages were affected by the main Scourian metamorphism, dating back
to about 2800 m.y.  If assemblage 1 is itself the reworking of older
materials then a very old history is represented in the Outer Isles.

Bowes and Hopgood (1973) have also recently outlined a framework for the
Lewisian of the Outer Hebrides.  They see the two complexes as an intimate
part of a crystalline basement representing a deep level of erosion below
cover rocks that are now gone.  They envisage a sequence of sediments on top
of volcanic lavas, pyroclastics and derived sediments, these volcanics passing down from acidic to andesitic to basaltic.  The true basement they see as
an anorthositic early crust.  This increasingly basic character with depth
is again a characteristic of the "greenstone belts" (and, by the way, also of
modern island-arc assemblages).

Before leaving these fascinating Outer Isles, mention must be made of one
other view regarding the "grey gneiss" complex.  Moorbath, Powell and Taylor
(1975) discuss isotope evidence for the age and origin of the complex. They
rule out the possibility that the gneisses represent the "reworking" of a
much older gneissic basement complex with anything like normal crustal Rb/Sr

ratios.  They further suggest that the precursors of the grey gneiss complex
were derived from upper mantle source regions not more than 100-200 m.y.
prior to the Scourian metamorphism.

The Lewisian of N.W. Scotland can therefore be seen to span a great length of
time.  It is the immediate foundation of N.W. Britain.  The question that
then arises is whether a foundation of the same immense age-span can`be seen
elsewhere in the British Isles.  The other areas of Precambrian in Britain
have already been considered in the previous chapter and formed part of the
Proto-Atlantic margins.  Many of these Precambrian remnants date back no
further than 1000 million years ago, and yet there must be a Lewisian-
correlated basement in England, Wales and Ireland.  The isotopic dating
(2475$\pm$180) by Max of gneisses in the Rosslare complex opens up the possibility
that other very old basement areas may be fortuitously revealed on the
British surface.  The Rosslare gneisses occur in association with the
Monian-type Cullenstown Group and this opens up the possibility of a Lewisian
basement at least very near to the surface in other Monian areas (South Irish
Sea to Anglesey).  Areas like Rushton, Primrose Hill and the Malverns in the
Welsh Borderland also spring to mind.  The end-Proterozoic dates obtained
from the Malvernian merely record an event there. After all, K/Ar dates of
500-600 m.y. have been obtained from Lower Proterozoic and even Archaean
rocks in N.W. France.  It is significant that two distinct episodes - the
Icartian with an age range of 2700-2550 m.y. and the Lihouan 2000-1900 m.y. -
have been detected in the Cherbourg Peninsula.  These accord with Scourian
and Laxfordian.


## THIS EARLY WORLD

The early Precambrian remnants now exposed in Britain are small and one needs
to go elsewhere, to the large Precambrian shields, to find out more about the
Earth's early history (and thereby of Britain's earliest history).  In
particular, one can call on the evidence obtained from Greenland, Scandinavia
and Canada, regions which were, for at least some times in the Precambrian,
continuous with one another.  There are close correlations in events between
the Lewisian and the Archaean-early Proterozoic rocks of these shields.  In
Finland, a sedimentary sequence immediately precedes the Svecokarelian
orogenic episode (1950-1900 m.y.).  That sedimentary sequence rests on rocks
dated at about 2800 m.y.  Bikerman, Bowes and Van Breeman have pointed out
that the correspondence of Precambrian chronology between Scotland and
Fennoscandia implies unified development in a major crustal segment at least
during the 2.8-1.5 b.y. period.  When extended to Greenland, though individual
peaks of tectonothermal activity do not quite correspond, nevertheless it can
be said that widespread orogenic activity characterised the eastern (Baltic
Shield-Scotland) and Greenland crustal segments between 2.0-1.7 b.y. ago
(Laxfordian).  In the Godthaab area of W.Greenland a 2800 m.y. deformation
resulting in granulites and high amphibolite facies metamorphism has been
recognised.  Significantly also it was followed (2600 m.y.) by granite and
pegmatite activity and thence by dolerite dyke intrusion.  There is here a
close Scourian comparison.  In Canada there was a major orogenic episode at
2750-2700 m.y. and another at 1900-1700 m.y.  To go back a stage further, the
rocks in W.Greenland can perhaps provide answers.  In this Godthaab area,
McGregor has deduced a pre-Scourian history.  The 2800 m.y. metamorphic event
was preceded by the Nuk intrusive phase (tonalites, granodiorites and
granites) and this in turn preceded by anorthosite emplacement.    In this
respect it is significant that, as Bowes and Hopgood have reminded, the

Scottish Lewisian could overlie an anorthositic crustal layer.  On the other
hand, later "underplating" could complicate the evolutionary order of this
sequence.  In W.Greenland McGregor's sequence of events goes back yet further,
back through thrusting, outpourings of basic volcanics and sediment deposition, basic dyke activity and ultimately to some of the oldest rocks
known, the Amitsoq Gneisses - granitic gneisses dated at 3750 m.y.    The
story does not even end there! Eighty miles N.E. of Godthaab at Isna, these
ancient gneisses contain large xenoliths of volcanics (suggesting a greenstone belt), sediments (including conglomerates with granitic boulders yield-
ing a 3800 m.y. date) and banded ironstones.  One immediately asks if all this
long story applies also to Scotland and does a similar great sequence underlie the Scourian?  Is there some correlation here with the metasediments of
South Harris in the Outer Isles?  On the other hand, was the Greenland crustal
nucleus there much earlier in time and was the Scottish crustal portion welded
on later?  In this event there may not be any Scottish age determinations
older than three billion years.

This question which has just been asked is but one of many asked of the
earlier Precambrian.  There are so many to ask:  What was the first crust
like?  Was it acid, basic or anorthosite in composition?  Was the early
crust continuous or was it patchy?  What was the earlier atmosphere like?
Were there oceans?  What are the "greenstone belts"?  Were there plate movements in Archaean times?  Were the "greenstone belts" related to plate subduction?  Did the early Earth suffer the same megaimpacts from outer space
as did the Moon and Mars?   The reader is particularly directed to the
excellent treatment of the problems of the Precambrian record in vol 2 Earth
History, Part I of Read and Watson's Introduction to Geology.   They favour
the probable derivation of the earth's atmosphere and hydrosphere by leakage
or outgassing from the deep interior.   Any initial atmosphere may well have
been lost.   They refer to Rubey's contention that the crystallisation of
a 40 km granitic shell could supply all the water in the oceans.  They point
out that early Proterozoic structural patterns appear to show continuity
over areas of considerable dimension.   The 2300 m.y. dyke swarms seem to
line up on a great circle from Canada to Greenland and Scotland.   By early
Proterozoic times, masses of continental crust of almost modern continent
standards were in existence.  These masses did move, as indicated by palaeomagnetic studies, relative to the magnetic pole but the masses moved as units
with no fragmentation and dispersal.   Mobile belts were developed within
and as part of the major crustal masses.  One gets the impression of a less
rigid outer crust than was to exist in later geological times or exists today.
The answer here may lie in (a) much greater radioactive thermal activity and
(b) a steeper geothermal gradient in the pre-Proterozoic Earth.   Fyfe has
suggested that the earlier heat production was twice or three times greater
than today.  The early Earth was in a thin granitic crust which was very
unstable.  Granitic activity was related to thermal anomalies in the mantle,
to localised loading and thickenings of the crust and to thrust tectonics.
There was faster melting at much shallower levels.  Archaean convection
cells were probably of smaller size.  Shackleton (1973) has suggested that
the characteristic metamorphic association of the Archaean, from greenschist
through amphibolite to granulite, indicates thermal gradients then in excess
of $30^{\circ}C$ per km as opposed to present day gradients in continental crust of
$10^{\circ}C$ per km.   He further suggests that the rate of radiogenic heat production
2700 m.y. ago was twice its present value and before that date was even
greater.  Ideas about the thickness and extent of early crust differ.  Fyfe
envisaged an early highly radioactive, easily fusible granite crust 5-10 km
thick but points out that any disturbance of the crust would result in

melting at its base producing magmatic activity thereby thickening that part
of the crust. Moorbath believes that large portions of Archaean sialic
crust were at least 25 km thick. With the higher radiogenic heat production
that crust would be heated and granitic melts would move upwards leaving
lower crustal levels as a kind of barren residue in the form of the high
grade metamorphic granulites we see in the earlier Scourian and pre-Scourian.
Moorbath believes also that there were particular times when the transfer of
material from mantle to crust were especially active. Some of these times
may well coincide with abrupt changes in continental movement as noted by
"hairpins" in the palaeomagnetically-deduced apparent polar wander paths.
Hairpins in the polar wander curve for North America, for example (see Read
and Watson, 1975, fig. 10.3), were at 2500, 1950, 1300 and 1100 m.y. These
can be correlated with high crustal activity (thermal and metamorphic) of
the Scourian, Laxfordian and Grenville phases. To some extent also there is
a cyclic pattern or timing in some of these hairpin figures. Runcorn has
noted the link between radioactive cluster dates and major tectonic activity
in shield areas.

Finally in this brief consideration of the early world (and thereby the
earliest "Britain") there are the problems of crustal thickening, meteoric
impacts and early plate movements. These problems may even be interrelated
as also is the problem of the early Precambrian "greenstone" belts.
Thickened crust could be the result of, for example, (a) increased magmatic
products from the mantle, the result perhaps of massive impacts from outside
or from periodic, perhaps cyclic thermal disturbances in the upper mantle;
(b) the collision of already existent continents; (c) plate tectonics, i.e.
sea-floor spreading and subduction. Recent studies of Mars show that most of
its southern hemisphere, the oldest Martian terrain, is an impact-scarred
surface representing early bombardment by objects of up to asteroidal size.
Continued internal activity after this impact phase resulted in the volcanic
flooding of vast areas, especially in the northern hemisphere and the surface
was cut by complex rift faulting. The great mare basalt floodings on the
Moon took place between 4.0 and 3.2 b.y. ago, after the great bombardment of
the Moon's highland region. The Earth must surely have suffered similar
early impacts with subsequent magma floods. Green (1972) has suggested that
the great Precambrian greenstone belts of Africa, W.Australia and Canada are
the equivalent of the lunar maria and highlands. The ultramafic komatiites
and the basic and andesitic lavas are often in irregularly arcuate or cuspate
synclinal areas irregularly interspersed by granitic or grandiorite gneiss.
Green attributes the more irregular diffuse relationship of outpourings to
impacted granitic shell in the case of the greenstone areas as being due to
the more mobile Earth retaining its "heat machine" much longer than the more
rapidly cooling Moon where the crater-flood relationship would be more
clear-cut.

Other opinions differ however and discount any extra-terrestrial causes. One
view is that as upper mantle basic or ultrabasic material welled up through
a thin granitic shell, outer covers of basic (and thereby denser) material on
top of lighter granitic crust would be an unstable situation and a churning
turnover would result with basic material being dragged down and remobilised
granitic material rising. This could explain the often very irregular pattern of basics to granite in these greenstone belts. Other views see the
volcanic assemblages as resting on gneissic basement. A body of opinion
discounts any large-scale plate motion in Archaean (or even Proterozoic) times.
Shackleton points out that the Pan African domains show no tectonic, structural stratigraphical or palaeomagnetic data to suggest large scale plate

motions, and no convincing ophiolites or sutures have been recognized within
these areas.  Talbot (1973) however compares the upward change within
greenstone volcanic piles (from ultramafic at the base to andesites,
and rhyolites) with the igneous assemblages of modern island arcs and ocean
floor.  Talbot discusses the possible models:  (1) the unstable denser on
lighter model with resultant "turnover" (discussed previously);  (2) thin
sialic crust with rift structures, i.e. incipient "spreading";  (3) sea-floor
spreading and subduction, and comes down in favour of the last theory.  Many
others have stressed the close comparison in upward (increasing calc-
alkaline) change between greenstone volcanic piles and modern island-arc
assemblages.  Dewey and Horsfield (1970) suggest that a relationship of
continents, island arcs and oceans has been determined by an ocean-based
plate mechanism for at least 3000 million years.  At first in the Precambrian
there were numerous and thinner ocean lithosphere plates and these controlled
the growth of continental crust in small nuclei  (a situation, perhaps, like
the S.W. Pacific today).  In this respect, it could then be possible that the
first earth crust was a basic one rather than a granite shell and that sialic
nucleii formed by plate tectonics as that basic shell was ruptured and moved
about.  Against this plate subduction theory, on the other hand, there remains the strong weight of opinion that (a) crustal masses may have moved as
units in earlier times and (b) that a major change in world tectonics took
place relatively late in Precambrian times.  The problem must be left there
as much more fact-finding has to be done.


## Suggested Further Reading

Bikerman, M., Bowes, D.R. & Van Breeman, O. 1975.  Rb-Sr whole rock isotopic
    studies of Lewisian metasediments and gneisses in the Loch Maree region,
    Ross-shire.  Jl. geol. Soc. Lond. 131, 237.
Bowes, D.R. & Hopgood, A.M. 1973.  Framework of the Precambrian Crystalline
    Complex of Northwestern Scotland.  In: Geochronology and Isotope Geology
    of Scotland: Field Guide and Reference.  (Ed. R. T. Pidgeon et al).  3rd
    European Congress of Geochronologists.
Craig, G.Y. (Ed.) 1965.  The Geology of Scotland.  Oliver and Boyd.
    Edinburgh & London.
Dewey, J.F. & Horsfield, B. 1970. Plate Tectonics, Orogeny and Continental
    Growth.  Nature, Lond.  225, 522.
Evans, C.R. & Lambert, R. St. J. 1974.  The Lewisian of Lochinver,
    Sutherland: the type area for the Inverian metamorphism.
    Jl.geol.Soc.Lond.  130, 125.
Green, D.H. 1972.  Magmatic activity as the major process in the chemical
    evolution of the Earth's crust and mantle.  Tectonophysics. 13, 47.
Moorbath, S., Powell, J.L. & Taylor, P.N. 1975.  Isotope evidence for the
    age and origin of the "grey gneiss" complex of the southern Outer Hebrides.
    Jl. geol. Soc. Lond. 131, 213.
Park, R.G. & Tarney, J. (Editors) 1973.  The Early Precambrian of Scotland
    and Related Rocks of Greenland.  University of Keele.
Read, H.H. & Watson, J. 1975.  Introduction to Geology. Vol 2.  Earth
    History, Part 1.  MacMillan Press. London and Basingstoke.
Shackleton, R.M. 1973.  Problems of the evolution of the continental crust.
    Phil. Trans. R. Soc. Lond. A273, 317.
Talbot, C.J. 1973.  A Plate Tectonic model for the Archaean crust.
    Phil. Trans. R. Soc. Lond. A273, 413.
Taylor, S.R. 1975.  Lunar Science : A Post-Apollo View.  Pergamon Press.
    Oxford.

## CHAPTER 4

# The Old Red Sandstone Continent

The Devonian Period spans some 50 million years of time. The Lower Devonian
occupied about half of this span, Middle and Upper Devonian times filled the
remainder. Important changes in the World picture took place by the beginning of Devonian times. Though the latitudinal position of Britain and of
N.E. America remained relatively unchanged (at about $25^o$ south of the
Equator), "Gondwanaland" (South America, Africa, India, Antarctica and Australia) had moved appreciably since Lower Ordovician times. The Africa-South
America "join" had in fact moved steadily across the South Pole so that this
pole changed its position from being over the West Sahara in the early
Ordovician to being at about the Argentine-S.W. Africa junction by the early
Devonian. As a result northernmost Africa moved from a polar position to be
only about $30^o$ south of the Equator. Spain and the Mediterranean area also
moved up into much lower (Southern Hemisphere) latitudes. At the same time,
some "swivelling" of the Gondwanaland fringe nearest to North America and
Northern Europe resulted in (a) a narrowing of the distance between North
America and South America but (b) a widening distance between Northern Europe
and Africa (with the Mediterranean region). The Devonian was possibly the
time when the Hercynian (Variscan, Rheic, Mid-European) Ocean might have been
at its widest. Whether it was an ocean or a vast spread of shallow irregular
seas is another matter, to which attention will be drawn later in this chapter.

Fig. 31 gives the approximate World picture for Devonian times. It will be
seen that the Proto-Atlantic Ocean has now closed and priority is given now
to the Mid-European and Urals "oceans". Most of the present day land areas
were (as previously) in the Southern Hemisphere (including now Australia,
which had in the meantime crossed the Equator). Only Asia (excepting India
and Arabia) found itself in the Northern Hemisphere. Even the western portion of North America had moved across to be south of the Equator. The North
Pole continued to lie in the West Pacific though a little further north than
previously. Waage and Storetvedt (1973) place it (on results from the Old
Red Sandstone of Caithness) at about $20^o$N $150^o$E. They point out, by the way,
that a discrepancy between this Caithness polar position and that obtained
from Norwegian rocks could indicate a 300 km sinistral displacement along
the Great Glen Fault of Scotland since the late Devonian.

The Devonian picture in Britain is the follow-up of the closing of the
Proto-Atlantic Ocean. Both sides of that ocean were now welded firmly together with the basin and marginal sediments rucked up into folded
mountainous areas. The sinking relics of oceanic crust were now gone forever
from view but their melt products intruded upwards to form the important
group of Caledonian (or "Newer") Granites. As stated by Reid and Watson,
most fall within the range 410-380 m.y. Thus some of them cross the basal
unconformity of the Old Red Sandstone. These late orogenic granites form
more than fifty separate bodies in Britain. They include both forcefully
emplaced plutons and ring complexes emplaced by "cauldron-subsidence (Glencoe and Ben Nevis, especially). The majority are sodic and granodioritic in
composition. Of the fifty British examples, about thirty are in Scotland

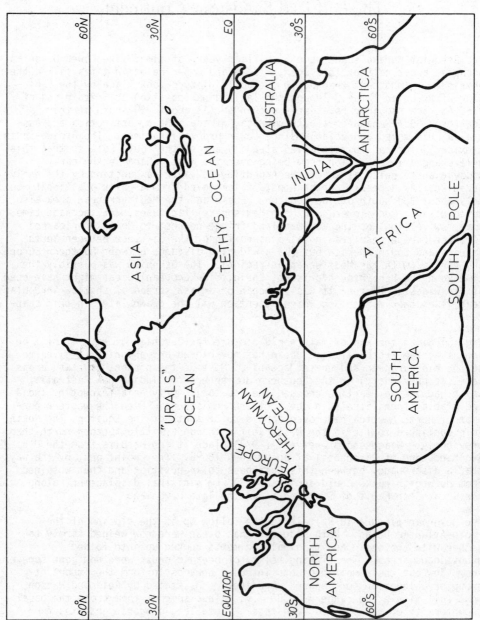

Fig. 31.  The Lower Devonian World (based on Smith, Briden and Drewry).

(see "The Geology of Scotland", fig. 7.7) and a large number of these are in that portion of the Highlands that lies between the Great Glen and Highland Boundary fractures.

The Glencoe and Ben Nevis activity was accompanied by caldera-subsidence and took place well into Lower Devonian times, being accompanied by outpourings of andesite and rhyolite, especially in the Etive and Lorne areas.  Roberts (1974) has suggested that the Glencoe caldera  was lop-sided with volcanic activity being concentrated on its north-eastern margin.  The caldera  was formed by flap-like subsidence of the floor and the underlying magma chamber was probably layered.  The andesites and rhyolites reach a total pile of almost 1200 m.

Further south, Caledonian plutons occur in the Southern Uplands and the Lake District.  In Ireland, the Leinster Granite is the largest mass in Britain. Bott (1974) has shown that an E-W negative Bouguer anomaly exists across the north and central Lake District and suggests that an underlying granite batholith extends to 7-10 km depth and links eastwards with the hidden Weardale Granite.  The latter, beneath the Alston Block, has been dated at $410\pm10$ m.y.  Dunham (1974) has described the discovery of another granite - the Wensleydale Granite - beneath the Askrigg Block.  The Rb/Sr date gives a $400\pm10$ reading but K/Ar data record also a 300 m.y. thermal event (perhaps connected with the Carboniferous Whin Sill activity).  Dunham concludes from strontium ratios that whereas the Weardale Granite originated in an upper mantle source region, the Wensleydale pluton resulted from refusion of existing ancient crustal material.  (It is tempting to see this difference as pointing to the possible position of the Lake District (Solway) subduction line).

With the closing of the Proto-Atlantic Ocean, Britain was now firmly welded to Scandinavia, Greenland and N.E. Canada to form a large "Old Red Sandstone Continent" (fig. 32).  Its southern fringe, as far as Britain was concerned, ran roughly E-W from just south of Ireland to the Bristol Channel and the Thames estuary (fig. 33).  To the north of this line lay a mountainous Britain but with already, by Lower Devonian times, much eroded areas and structurally controlled basins or troughs, the most notable being the Midland Valley of Scotland.  The highest relief probably lay in the northwest of Scotland and (for much of Lower Devonian times) in the Grampian Highlands. Parts of the Southern Uplands too were rugged.  Though fault-controlled to a considerable extent, the form of the Mid-Scottish basin did not exactly conform to the modern Midland Valley and extensions of the depositional area occurred in the Cheviot and Fort William districts.  Volcanic activity was occurring on an intense scale in Lower Devonian times in the Midland Valley, particularly in the Ochil-Sidlaw area, and again in the two extensions. Further south in Britain, volcanic activity in the land areas was negligible, with the exception of some Irish areas.

The southern edge of the Old Red Sandstone Continent was of much more reduced relief and wide low embayments, practically at sea level extended northwards into it.  Many of these embayments were in fact relics of late Silurian waters and in some, for example parts of the Welsh Borderland, deposition continued without interruption from Ludlow into Downtonian times, that is across the previously accepted Siluro-Devonian boundary (recent deliberations may well place this considerably higher in the Lower Devonian sequence thereby returning the Downtonian to its original Upper Silurian position).

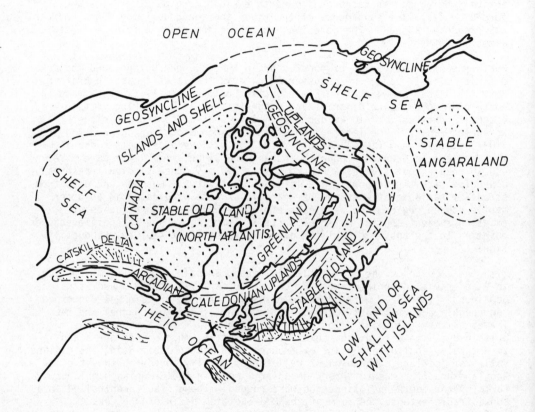

Fig. 32.  (Present) Northern Hemisphere reconstruction of
Devonian times (Dineley, D.L. 1973, fig. 8).

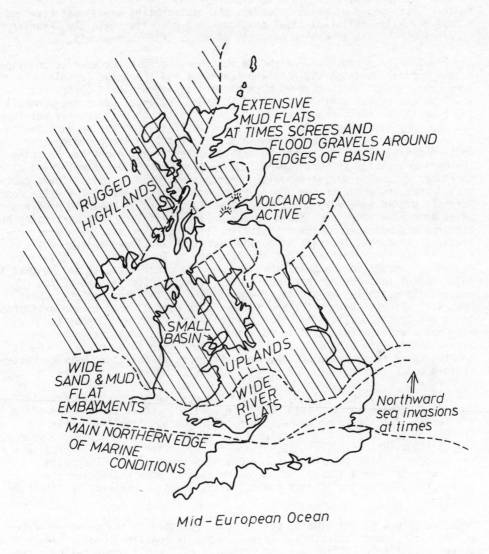

Fig. 33.  General palaeogeography of Devonian times.

As a result of the low-lying character of portions of the southern border of
the landmass, the Lower Devonian often commences with thicknesses of
relatively fine-grained sediments, marine at first but becoming increasingly
less marine upwards. This is true especially of the Welsh Border areas.
Further west into S.W. Wales, however, the contributing areas must have had
a more irregular relief as basal conglomerates or grits begin the Devonian
succession. This is true also of parts of the S.Irish embayments.

The E-W line from south of Ireland to the Thames marked the average position
of the boundary between the northern land area and the Rheic (or Mid-
European) "Ocean" which covered much of Central and Southern Europe and
which spread as far north as S.W. and Southern England. Devon and Cornwall
lay beneath the northernmost waters of this sea though whether this was true
for the very beginning of Devonian times is difficult to say because the
Siluro-Devonian base is never seen in S.W. England.   The boundary line
between sea and land fluctuated through Devonian times, sometimes bringing
non-marine deposits to North Devon (the Hangman Grits, for example) and
sometimes spreading marine deposition northwards into the coastal embayments.
Marine horizons occur in the Upper Devonian of South Wales.   The Willesden
Borehole proved some very thick Upper Devonian strata. Lower to Middle
Devonian marine rocks were found in the Witney Borehole. More surprising has
been the discovery of marine deposits in Cambridgeshire and, even more surprising, the recent location of marine Devonian limestones in the Argyll
Field of the North Sea (at about the latitude of Edinburgh).    This must
represent a major marine break-through. On the other hand, it may be that a
major indent into the northern continent always existed in some of the
present-day North Sea area.  It is significant in this respect that Upper
Devonian palaeoslopes in Fife and Kinross have been deduced to dip eastwards.
The great Orcadian "lake" in which the Caithness Flags were deposited may
have had a wide eastward and southward extent in the North Sea area and
marine floods into part of this embayment could have taken place.   One is
tempted to at least think of very early attempts at splitting between East-
Britain and Scandinavia.

There are signs also of rising areas in the seas to the south of Britain.
Some of the thick spreads into Kerry, in Ireland, appear to have come from
the south.  The spread of Gramscatho-type greywackes into the southern
fringes of Cornubia in Mid-Devonian times again hints at a southerly source.
Renouf (1974) has (his figs. 3E, 3F and 3G) suggested a rising source area in
the Western English Channel in Lower Devonian times.  This "Domnonaea"
supplied debris to Brittany, on the other side.   Some L.Devonian conglomerates at Roseland (Lizard Peninsula) have debris derived from the same
barrier.

"Domnonaea" may be the simple, but correct, solution to the palaeogeography
of what was between Britain and France.  Some geologists do not, however,
accept this simple explanation and suggest that a whole ocean has been sub-
ducted from our gaze in this Channel area.  Bromley (1975) suggests that in
the relatively undisturbed eastern part of the Lizard Complex there is a
simple ocean crust-type succession from ultrabasics into gabbros which are
overlain by the root-zone of a sheeted dyke complex.   Burrett (1972)
postulated a "lost ocean" between Cornubia, the Ardennes, Rhenisches
Schiefergebirge and Harz on the northern side and Brittany, Vosges, Thuringia
on the southern side.  This ocean picture by Burrett is shown in fig. 34.
Volcanic outbursts in Cornubia, the Rhine and the Harz were then connected

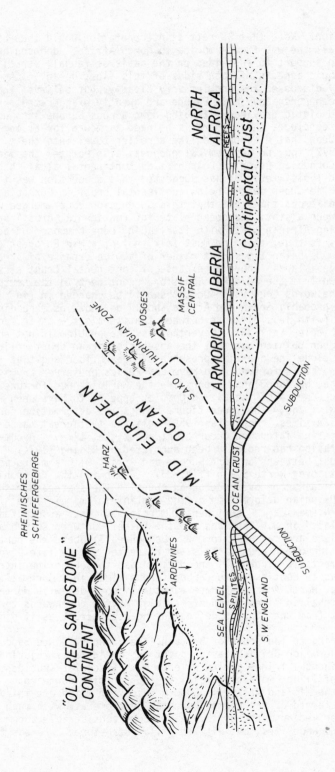

Fig. 34. Diagrammatic view of Europe in late Middle to early Upper Devonian times (after Burrett, 1972).

with early subduction.  Note that Burrett's interpretation would then have
continental crust all the way from Armorica to North Africa.  Johnson has
more recently given support to this idea on the basis of faunal, stratigraphical and tectonic aspects of both sides of this "lost ocean".   Ager
(1975) has recently discussed the problem very clearly, but believes that
the contrasts between north and south Europe are greatly exaggerated.   He
sees a single broad biotic province extending down across Europe into north-
west Africa.  Ager concludes that "there is no need to mourn for a long-lost
central European ocean that disappeared like a roller-towel into the
mantle".  Floyd (1972) too, on geochemical grounds, also refutes the idea
and explained the Cornubian Devonian volcanics as continental alkali basalts
rather than oceanic tholeiites.  He has suggested that subduction was taking
place deep down in the upper mantle below continental crust in Cornubia in
Devonian and Carboniferous times but that this subduction zone emerged well
to the south - between a static European plate (of continental crust) and a
north-easterly moving African plate with its leading edge composed of oceanic
crust.  If all this is true, then one must talk, not of a Mid-European (or
Hercynian) Ocean in Devonian times, but rather of a wide expanse of irregular
sea (with many large islands) above an expanse of continental crust stretching all the way down from the "Old Red Sandstone" continent to the north.
In this respect, the words "Hercynian Ocean" should be changed in fig. 31 to
Hercynian Seas and probably Spain and Africa should be shown as more directly
facing N.W. Europe (with less distance between the two sides - as in fact in
fig. 32, drawn by Dineley in 1973).  Nevertheless, the problem is not an
easy one.  It has been pointed out that the absence of ocean crust relics do
not necessarily preclude the one-time presence of ocean floor and that in
complete subduction, it is fortunate to have any relics preserved anyway.
Anglesey, Ballantrae, Newfoundland, Cyprus are the World's good fortune.  In
Cornubia the Kennack Gneiss of the Lizard Complex appears to have been metamorphosed 350-390 m.y. ago.  The lower figure brings the deformation forward
to very late Devonian times.  Hendriks (1959) believes that great nappe
formation and serpentine intrusion occurred at about that time.   A com-
promising interpretation has recently been submitted by Riding (1974).  He
envisaged ocean crust between Northern Europe and Africa with a micro-plate
(or micro-continent) over North Spain (fig. 35).  Following the Caledonian
Orogeny, North America-Europe overrode ocean crust to the southeast resulting
in Middle to late Devonian deformation of the continental margin (the
Acadian Orogeny in N.America).   Continental margin sediments associated with
basic lavas accumulated in S.W. England and the Rhine.  Between Cornubia
and Brittany an island arc may have formed ("Domnonaea"?) above the northward
sloping and subducting ocean crust.  This arc retreated in the late Devonian
deforming the sediments in S.Cornwall and giving rise to serpentine intrusion.
This deformation was to continue into Carboniferous times in S.Cornwall.
According to Riding, North America-Europe eventually collided with North
Spain in later Westphalian times, with Africa also moving northwards towards
the micro-continent from the other side, devouring ocean crust as it did so.

Whatever was happening to the south of Britain, the known "Old Red Sandstone"
accumulations in the major British basins point to considerable rejuvinating
pulses on the continental fringe giving rise to both irregular subsidences
and fairly rapid infilling, more especially in the Scottish downwarps.  The
Lower Devonian in the Midland Valley is 6000 m thick, of which one third in
Kincardineshire is formed of the coarse Dunnotar Conglomerate with much
Dalradian debris derived from the north.  Above come considerable accumulations of basalts and andesites, but with no pillow structure.  Whereas these

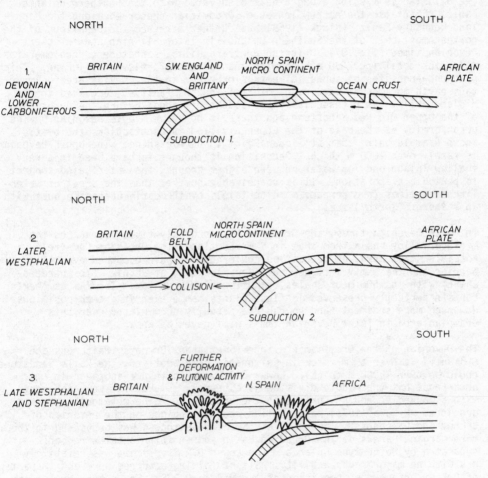

Fig. 35.  Devonian and Carboniferous plate movements in
Western Europe (after Riding, 1974).
Horizontal shading - continental crust;  inclined
shading - ocean crust

conglomerates derived their debris from northern Dalradian sources, the
Hagshaw rudites of the southern side of the Midland Valley were carried from
the Southern Uplands.   George (1965) has reconstructed the floor on which
the Lower O.R.S. of Central Scotland was deposited (fig. 36).   It shows that
the Silurian is missing along a belt associated with the Southern Uplands
Fault.   Note also the narrow lenses of Ordovician preserved along the Highland Boundary Fault.   Bluck (1969) has further stressed the behaviour of the
northernmost belt of the Southern Uplands in later Silurian and in Lower
Devonian times (fig. 37).    In the Pentland Hills, sandstones, conglomerates,
volcanics totalling 2500 m rest with angular unconformity on Silurian.   Thick
pyroxene-andesites of subaerial extrusion occur in the Cheviot area.   Important earth movements, involving both folding and faulting, occurred in the
Midland Valley basin (especially its southern fringe) before the deposition
of the Upper Old Red Sandstone and there is no known Middle Devonian.   The
unconformity at the base of the upper division is spectacular, the best
known example being that at Siccar Point in Berwickshire.   The Upper Devonian
is rarely over 1000 m thick.   Deposition is thought to have been in a very
shallow inland body of water between higher ground, the water being subject
to periodic evaporation.   It is conceivable however that the sheet of water
linked at times (and perhaps more certainly by the early Carboniferous) with
an eastern ("North") sea.

In Lower Devonian times, the Grampian mass probably firmly separated the
Midland Valley basin from that in N.E. Scotland (the Orcadian Cuvette).
Possible small areas of deposition could however have occurred over this
barrier and these may include the famous locality of Rhinie (in Aberdeen-
shire) with its 500 m of shales, sandstones and rudites.   The famous cherts
contain perfectly preserved leafless plants, some even in growth positions.
How much more sediment (and volcanics) actually accumulated over this
Grampian area in later Devonian times will never be known.

The Devonian of the Orcadian Province (Shetlands, Orkneys, Caithness and
Cromarty) is over 6 km thick in Caithness.  (In Shetland, the Walls Sandstone
could be over 12000 m thick).  Lower O.R.S. deposits are sporadic in
occurrence but the middle division is extremely thick.  The Devonian rests
with a marked irregular surface on the underlying Moine metamorphics or on
granites.  Mykura (1975) has supplied recent findings on the Devonian of
Orkney and Caithness.  The Lower O.R.S. of S.Caithness was deposited in the
more marginal areas of the Orcadian Basin and in intermontaine regions
separated by Moine-type uplands.  The Lower O.R.S. climate was relatively
dry with no major river activity.  Slight folding occurred here and in parts
of Orkney before the deposition of the Middle O.R.S.  Scree-type deposition
around the Moinian margins of the south pass laterally into yellow aeolian
sands in Orkney.  Middle O.R.S. basal fan deposits pass upwards and laterally
northwards in Caithness into lacustrine deposits.  The thick Caithness Flag-
stone Series (5000 m) was deposited in a large shallow lake.  This sequence
represents repeated lacustrine cyclothems similar to those seen at the present
time in the Caspian Sea.  The cyclothem comprises a basal dark fissile limestone, then dark flags followed by current-bedded sandstone and lastly mudstones with dessication cracks and ripple marks.  Fish remains occur mostly
in the limestone unit.  The best known Achanarras Limestone cycle was a thick
one, deposited when the lake was its deepest and widest extent.   Its
equivalent is found in Orkney.  Mykura has drawn attention to the "Upper
O.R.S." Hoy Sandstone of Orkney.  He points out that K/Ar dating gives an

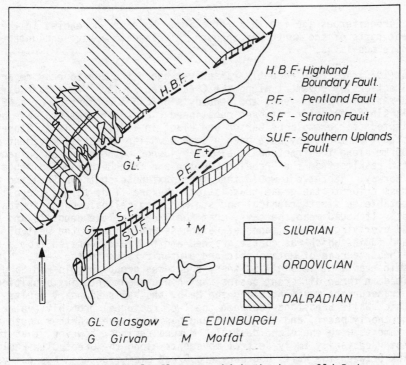

Fig. 36. Probable floor on which the Lower Old Red Sandstone was deposited (after George, 1965).

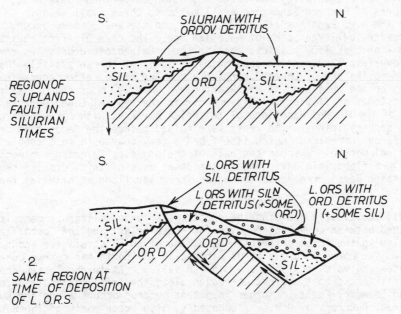

Fig. 37. Siluro-Devonian Palaeogeology of the Southern Uplands Fault area (after Bluck, 1969).

Upper Carboniferous age for this formation. Moreover boreholes in the
adjacent parts of the North Sea through this sandstone have encountered Upper
Carboniferous lavas.

Mykura (1975) has also recently raised yet again the controversy regarding
the directions of movement along the Great Glen Fault. Views have differed
considerably in recent years and Kennedy's original contention of a 65 mile
(total) sinistral shift has been questioned. Dextral movements have been
substituted (for example by Garson and Plant in 1972). In 1974, Storetvedt
suggested that there may have been a late Devonian sinistral movement of
200-300 km along the fracture zone, his evidence being based on palaeomagnetic results from the Orcadian Cuvette and from Devonian sediments in
South Norway. Storetvedt would therefore juxtapose southern Shetland against
Caithness prior to the great translation. Mykura finds this juxtaposition
unacceptable on stratigraphical and sedimentological-palaeogeographical
grounds. It would mean, he says, that the Orcadian-lake deposits and fine-
grained overlying fluvial sands of Caithness would lie to the west of a
Shetland facies which was coarse-grained and fluvially derived from a nearby
mountainous terrain of metamorphic and plutonic rocks to the west. Mykura
claims instead that in Shetland there are three groups of Old Red Sandstone,
deposited in three different basins, but now brought into juxtaposition by
major transcurrent movement along the Melby and Walls Boundary faults. The
very thick Walls Sandstone (between the two fractures) probably evolved in a
more northerly basin, and the E.Shetland basin was even further north again.
The movements were therefore dextral. Mykura reminds, however, that this
does not necessarily imply dextral movements along the Great Glen Fault as
the link between the Walls Boundary and Great Glen fractures has not yet been
proved.

Between the Midland Valley basin and the Welsh Borderland-Powys area,
occurrences of Old Red Sandstone are limited to three small areas. On Cross
Fell in the N.W. Pennines and at Ullswater, to the west, there occur
conglomerates believed to be of Devonian age. The Lake District rudites,
which may be over 1500 m thick, contain Lower Palaeozoic debris and are themselves overlain by Dinantian carbonates. It is of course possible that the
coarse deposits are of Lower Dinantian age. The third relic occurs in N.E.
Anglesey on the west side of Red Wharf Bay. Here 500 m of red sandstones
and siltstones of fluviatile origin occur in the typical Lower Devonian cyclic
fashion of the Old Red Sandstone of the Welsh Borders, with cornstones con-
cluding each cyclothem. The rocks were folded before the deposition of the
Carboniferous Limestone (which itself has a very coarse (almost cobble-grade)
basal conglomerate. Whereas the bulk of the Anglesey Devonian sediments
formed in a floodplain environment, the basal beds of this Devonian sequence
(the Bodafon Beds) were deposited on an irregular floor as alluvial fan
deposits.

These difficulties in deciding the exact age of unfossiliferous deposits
sandwiched between Lower Palaeozoics (or Precambrian) and the Carboniferous
apply also to the Irish outcrops. No marine Devonian strata are known in
Ireland. The Lower O.R.S. area of Fintona in the north has coarse conglomerates and volcanics like the Midland Valley. As in the latter area,
these Irish rocks abut against Dalradian along the Irish continuation of the
Highland Boundary Fault. Another important outcrop occurs along the Curlew
Anticline. Outcrops of O.R.S. (supposedly) also occur north of the Boundary
Fault, at Cushendall and Lough Swilly and, more importantly, in Mayo.

Further south, in the Dingle Peninsula, a Silurian sequence seems to pass up conformably into the fluvio-lacustrine Dingle Beds (2700 m thick) which could be of Downtonian age. These grey to purple sandstones and thick conglomeratic wedges were, with the Silurian, folded and even overfolded in a late Caledonian episode prior to the deposition of the Upper Devonian. This higher Devonian sequence, 1600 m thick, begins with the thick Inch Conglomerate made up of metamorphics derived from a southerly source. This elevation could have been still to the north of the Munster basin of deposition as an appreciable northward carriage of S.W. Ireland could have occurred along the fault zone of the Dingle Bay-Dungarvan "front". The (Upper only?) Devonian succession thickens appreciably south of Dingle Bay being 4000 m thick in the McGillicuddy's Reeks and as much as 7000 m in the Caha Mountains. The red, grey and green rocks are of fluvial origin. Further east, the deformed 3000 m sequence of the Comeragh Mountains could, like the Dingle Beds, belong to the pre-orogenic Devonian.

The Anglo-Welsh cuvette was one of the major "embayments" in the (British) southern border of the Old Red Sandstone Continent. The preserved outcrops of the sediments laid down in this important basin range from the West Midlands through the Welsh Borderland then southwestwards into Powys and South Dyfed (that is, Carmarthenshire and Pembrokeshire) with other extensions into Gwent, Glamorgan, Avon and Somerset. In the West Midlands and the Borderland the Devonian succession is 1400 m thick. This swells up to 2300 m in the Brecon Beacons and in Carmarthenshire. In S.Pembrokeshire, the total thickness is 1400 m but this could be trebled in the area north of the Milford Haven if the "traditional" thickness of 3300 is believed for the Cosheston Beds (recent work, still in progress, makes it likely that the figure is excessive). On the south and east rims of the South Wales Coalfield the sequences are thinner, as they are in the Mendips (only Upper Devonian here) and the Bristol-Clevedon districts (800 m).

The basin is the one in which the accepted stages of the continental facies ("Old Red Sandstone") were established. The stage names of the Lower Devonian (Downtonian, Dittonian and Breconian) and of the Upper Devonian (Farlovian) are all derived from localities in Shropshire or Breconshire. Middle Devonian rocks in the basin have not been proved, though the Ridgeway Conglomerate of South Pembrokeshire could be at least in part, of this age. Walmsley (1974) has discussed the problem of placing the new international position (base of the Monograptus uniformis zone) for the Devonian base within the Anglo-Welsh succession. It could be at the Downtonian-Dittonian junction, a boundary "where brackish intertidal sediments give way to fluviatile beds whether these begin with limestone or sandstones" (Allen and Tarlo, 1963).

In the Welsh Borderland, the Downtonian follows conformably on the Ludlovian. The base is the famous Ludlow Bone Bed, formed of vertebrate and shell debris with phosphatic nodules. The bed represents a time when terrigenous material was in short supply. According to Allen (1974) it would be a strand-line deposit formed during the transgression of a shelf area. The Downton Castle Sandstone (above the bone bed) represents littoral sand bodies of either a delta or a coastal plain. It may be the filling of a marine infilled gulf that was deepest in mid-Wales. The overlying Temeside Shales were deposited in broad mudflats associated with river mouths. Traced southwestwards the base of the Devonian oversteps Ludlovian on to Wenlockian and finally on to Ordovician near Carmarthen. At first the base is formed by the

very micaceous Tilestones (or Long Quarry Beds).  These give way westwards to
the more massive Green Beds.  South of Haverfordwest these basal beds are
conglomeratic and step down on to a Precambrian igneous basement.

The remainder of the Downtonian in the Welsh Borderland and West Midlands is
formed of 300-450 m of red "marls" (actually siltstones) with subordinate
sandstones and calcareous concretions.  The siltstones are associated with
thin channel-filling intraformational conglomerates.  Calcareous concretions
increase particularly towards the base of the Dittonian forming almost concretionary limestones.  The sandstones were (according to Allen) formed in
laterally wandering channels.  In the lower siltstones, the sandstones could
represent the tidal lower reaches of rivers but the higher arenites were
river channel deposits.  The siltstones represent tidal mud flats or river
floodplain deposits.  The flats were frequently exposed to atmospheric
processes (giving concretions and producing suncracks).

In the overlying Dittonian division, thick red siltstones give way to a more
regular alternation of sandstones, siltstones, concretionary limestones and
intraformational conglomerates.  The sequence is markedly cyclic and excellent cyclothems can be seen on some of the road sections through Hereford-
shire and Gwent (for example on the M.50 near Ross).  The best limestones
occur at the base ("Psammosteus" Limestone) and at the top (Abdon Limestone
of the Clee Hills) of the Dittonian.  Allen says the concretionary limestones
are like caliches.  These pedogenic carbonates require long periods (10,000
years) of exposure for formation.  The cyclic sequences may be the result of
base level fluctuations in combination with river avulsion.  Overall, the
Dittonian sequences were formed on a vast alluvial plain extending southwards
to the Devonian sea.  The caliches could be on slightly elevated interfluvial
areas.  Near the top of the  Dittonian in the Cardiff-Risca area there occurs
the Llanishen Conglomerate, conglomerate layers up to 5 m thick interbedded
with  sandstones and siltstones.  A persistent massive calcrete - the
Ruperra Limestone - occurs within the conglomerate formation.  The conglomerate clasts are mostly angular and include acid volcanics (from the
Mendips or Tortworth?), pink quartzite (with U.Llandovery fossils) and
porphyries.   Allen (1975) has given a detailed description of the Llanishen
Conglomerate and shows that a nearby southern source is most likely, perhaps
from a westerly extension (in Devonian times) of early Silurian volcanics
and sediments from Tortworth and the Mendips.  Allen revives the possibility
of Bristol Channel barriers and raises the question of the early Devonian
Dartmouth Beds of Devon having been formed in a separate basin from the Lower
Old Red Sandstone of South Wales.

Breconian rocks are particularly well developed in the county of the same
name and form the great north-facing scarps of the Brecon Beacons and Fans,
and the Black Mountains of Monmouthshire (now Gwent).  In the Beacons area,
the Breconian is made up of 300 m of flaggy Senni Beds and 450 m of the more
sandy Brownstones.  In the Clee Hills, the Breconian Woodbank Series, like
the Brownstones, is of alluvial origin.  Allen suggests that the alluvial
plains were now being crossed by more steeply sloping and less strongly
meandering rivers than existed in Dittonian times.  The equivalent rocks in
the area immediately north of the Milford Haven are the reputably-thick
Cosheston Beds, said to be over 3,000 m thick.  These rocks occur on the
north side of the important Ritec Fault.  South of that E-W fracture they are
absent, being replaced by the Ridgeway Conglomerate (up to 350 m thick)
which could be of Middle (or even higher) Devonian age.  This formation

contains quartzite pebbles with Lower Palaeozoic fossils and the clasts were transported from a southerly source. Allen believes this facies is like a playa-basin and the alluvial-fan sediments of modern semi-arid regions.

J.R.L. Allen has made a considerable contribution to our knowledge of the Devonian sequences and palaeo-environments of the Anglo-Welsh cuvette. In particular he has shown that the bulk of the Old Red Sandstone was derived from the north. His study of exotic pebbles from the Brownstones of Hereford-shire and Worcestershire (1974) indicates sources in Wales of Precambrian, Ordovician, Silurian, and early O.R.S. rocks. By late Lower Devonian times, denudation in the Welsh region had already led to the partial removal of early Devonian and Silurian rocks. Moreover this pebble assemblage is not like that of the Llanishen or of the Ridgeway conglomerates (and they are not like each other).

The Upper Old Red Sandstone in the Anglo-Welsh Basin is very much thinner. It must rest unconformably on lower rocks as the middle division is almost everywhere absent. In Shropshire there is visible discordance and in the Breconshire Fans, a small difference in tilt can be detected, but elsewhere in South Wales the detection of unconformity is sometimes not easy. Near Portishead, the Upper O.R.S. rest with obvious unconformity on more steeply dipping Breconian sandstones. Allen (1974, fig. 10) has reconstructed the palaeogeology of the Farlovian floor. Later structures such as the Usk Anticline, Forest of Dean Syncline, Malvern Axis and the Benton Fault were already being anticipated in the pre-Farlovian earth-movements. The Farlovian sediments are themselves very variable in character from region to region. Conglomerates characterize the southern and eastern rims of the South Wales Coalfield. The red gritty Plateau Beds on the north side of the Coalfield have yielded marine horizons with faunas including Cyrtospirifer verneuili. These temporary marine advances were followed, according to Allen, by a retreat and emergence with the fluvial Grey Grits ultimately forming the third episode of sedimentation in Breconshire. These grits, like the equivalent partly-marine Skrinkle Sandstone of South Pembrokeshire, grade up into the Carboniferous. The Plateau Beds have been compared by Allen to sediments accumulating in a coastal barrier complex. South Wales and the Welsh Borderland was part of an extensive coastal plain during Upper Devonian times. That plain was transgressed twice from the south by the sea.

In S.W. England, the Devonian occurs in three main areas. In the north are the outcrops on Exmoor and the Quantocks. The third area, by far the largest, occupies much of Cornwall (south of Boscastle and Launceston) and South Devon (south of a line from Tavistock to Newton Abbot, with a small inlier to the north at Chudleigh). This southern area is broken by the Lizard-Dodman-Start thrust zones, on the coastal fringe, and by the great Variscan granites (Dartmoor, Bodmin, etc.). The base of the Devonian is nowhere seen in S.W. England. In North Devon, the oldest exposed Devonian is only just down in the Lower Devonian (Upper Emsian Stage); in South Devon the mainly non-marine Dartmouth Beds are Dittonian or Siegenian at the oldest. The Lizard Complex and its northern zone of complex geology may hold earlier Devonian representatives. In the Roseland area some Gedinnian shales, limestones and spilites seem to be present. The Start Schists could just possibly be their (now metamorphosed) equivalents. In southernmost Cornwall some Lower Devonian appears to be represented by the Mylor Beds, followed northwards by the Middle Devonian Gramscatho Series (slates and greywackes of flysch origin with

northward transported debris). The Dartmouth Slates occupy much of the axial
zone of the main Watergate Anticlinorium, a major E-W structure running from
Newquay to Dartmouth. Above these fluvio-deltaic, green and purple sandstones and siltstones, with pyroclastics, come the marine Meadfoot Beds and
their lateral equivalents, the Staddon Grits and Newquay Slates. This com-
pletes the Lower Devonian assemblage. The Middle Devonian of South Devon is
a thick group of muddy sediments (now slates) with intercalated, lenticular
sheets and masses (often of considerable size) of limestone and volcanics.
The carbonate masses are richly fossiliferous in stromatoporoids, corals,
brachiopods and polyzoa. The best limestone lenses occur at Brixham,
Plymouth, Torquay and Chudleigh. The Plymouth Limestone is almost 400 m
thick and dominates the scenery of that great harbour. In Cornwall, to the
west, the Trevose Slates and underlying pillow lavas of the Padstow area
make up much of the Middle Devonian. The Upper Devonian then occupies much
of the remainder of this north Cornish coast as far as Tintagel and Boscastle.
Structures are complex and there are great thrusts running almost parallel
with the coast. Pillow lavas occur at Pentire followed by slates and conglomerates. Volcanics occur widely also in the Tintagel area. Nearer
Torquay, Middle Devonian vulcanicity is represented by the Ashprington
Volcanic Series. The Upper Devonian of South Devon is made up largely of
ostracod-bearing slates with thinner cephalopod-bearing limestones and shales
taking their place along the northern edge of the Devonian tract (for
example, at Chudleigh).

In North Devon and West Somerset, the oldest Devonian exposed occurs at
Lynmouth. These Lynton Beds are marine but are followed by the thick Hangman Grits (1300 m thick and of mainly Eifelian age) representing the lower
of two "Old Red Sandstone" incursions from the north. It is significant that
this southwardmost advance of continental conditions corresponds with an
absence of Middle Old Red Sandstone in most of the Anglo-Welsh cuvette. The
second southward incursion formed the Pickwell Down Sandstone, of the same
order of thickness, in the Upper Devonian. Between these two O.R.S. facies
occur the marine Ilfracombe Beds (with several limestone horizons) and the
cleaved Morte Slates. The highest Devonian strata, the Pilton Beds, continue
without a break into the Lower Carboniferous.


Suggested Further Reading

Ager, D.V. 1975. The geological evolution of Europe. Proc. Geol. Ass. Lond.
   86, 127.
Allen, J.R.L. 1974. The Devonian Rocks of Wales and the Welsh Borderland.
   In: The Upper Palaeozoic and Post-Palaeozoic Rocks of Wales,
   (Ed. T.R. Owen). Univ. of Wales Press, Cardiff.
Allen, J.R.L. 1974. Source rocks of the Lower O.R.S.:Exotic Pebbles from
   the Brownstones, Ross-on-Wye, Hereford and Worcester. Proc. Geol. Ass.
   Lond. 85, 493.
Allen, J.R.L. 1975. Source rocks of the Lower O.R.S.:Llanishen Conglomerate
   of the Cardiff area, South Wales. Proc. Geol. Ass. Lond. 86, 63.
Burrett, C.F. 1972. Plate Tectonics and the Hercynian orogeny.
   Nature, Lond. 239, 155.
Dineley, D.L. 1973. Earth's Voyage through Time. Hart-Davies, MacGibbon, Lond.
Dineley, D.L. 1973. The Fortunes of the Early Vertebrates. Geology. 5, 2.
Floyd, P.A. 1972. Geochemistry, origin and tectonic environment of the basic
   and acidic rocks of Cornubia, England. Proc. Geol. Ass. Lond. 83, 385.

Garson, M.S. and Plant, J. 1972.  Possible dextral movements on the Great
  Glen and Minch Faults in Scotland.  Nature, Lond. 240, 31.
Mercy, E.L.P. 1965.  Caledonian Igneous Activity.  In: The Geology of
  Scotland, (Ed. G.Y. Craig).  Oliver and Boyd, Edinburgh and London.
Mykura, W. 1975.  Orcadian Geology - O.R.S. of Orkney and Caithness.
  "Eclogos".  Univ. of St. Andrews. 1, 48.
Rayner, D.H. 1967.  The Stratigraphy of the British Isles.  Cambridge
  University Press, Cambridge.
Read, H.H. and Watson, J. 1975.  Introduction to Geology. Vol. 2. Earth
  History, Part 1.  MacMillan Press. London and Basingstoke.
Renouf, J.T. 1974.  The Proterozoic and Palaeozoic Development of the
  Armorican and Cornubian Provinces.  Proc. Ussher Soc. 3, 6.
Riding, R. 1974.  Model of the Hercynian Foldbelt.  Earth and Planetary
  Science Letters.  24, 125.
Smith, A.G., Briden, J.C. and Drewry, G.E. 1973.  Phanerozoic World Maps.
  In: Organisms and continents through time (Ed. N.F. Hughes).
  Palaeont. Ass. spec. papers.
Storetvedt, K.M. 1974.  A possible large-scale sinistral displacement along
  the Great Glen Fault in Scotland.  Geol. Mag. 111, 23.
Waterston, C.D. 1965.  Old Red Sandstone.  In: The Geology of Scotland,
  (Ed. G.Y. Craig).  Oliver and Boyd, Edinburgh and London.

## CHAPTER 5

# "From Coral Seas to Mountain Chains"

The Carboniferous Period, lasting 65 million years, was an important one, from the British point of view, for several reasons. Firstly it was the time when many of Britain's important mineral resources - limestone, minerals, refractories and of course coal - were formed. Secondly, it was a time of great changes. The period began with the waters of the southern seas invading northwards into many basins and depressions formed on the Old Red Sandstone continent. These waters, often clear at first, were to be later polluted by sandy and muddy debris as upland areas became rejuvenated. Vast muddy and sandy spreads formed at almost sea level and became densely forested at times in the now truly equatorial climate which Britain was experiencing. That sea level was to oscillate on a rhythmic scale for a considerable time and for reasons which are complex. The movements may have been the forerunners of a gathering storm which beginning far to the south gradually spread north and converted the British area once more into part of a continental area of diverse and often considerable relief. Thirdly, the Carboniferous Period was the one during which "Britain" crossed the palaeo-equator, changing its latitude from about $15^{o}$S in Lower Carboniferous times to $10^{o}$N in the early Permian. The British area had been situated in the Southern Hemisphere for a long time previous to this. From late Carboniferous times onwards it was to remain now in the Northern Hemisphere.

The South Pole remained in much the same place from Devonian into Carboniferous times, being situated still in the boundary region of Argentina-South Africa. These two continents in fact remained fairly still. The North Pole moved somewhat north over the West Pacific. There appears to have been some appreciable rotary swivelling however of the North American-North European areas with a substantial narrowing of the ocean between the eastern seaboard of North America and West Africa (the "Proto-South Atlantic"). Britain too would have been involved in this "eastward" and somewhat "northward" drift bringing Northern Europe nearer to the kind of relationship which it has with the African continent and Southern Europe at the present time. It follows from the northward component of drift of North American and Northern Europe that areas such as N.Greenland, N.W. Canada and northern Scandinavia (all previously very close to the Equator) had now drifted well into the Northern Hemisphere. These fundamental changes in the positioning of N.W. Europe relative to the Equator and relative to other continental plates from Devonian to U.Carboniferous times could, at least in part, account for the important changes in climate from the caliche-forming arid phases of the Devonian to the wet, swampy-forest conditions of the coal-forming phases.

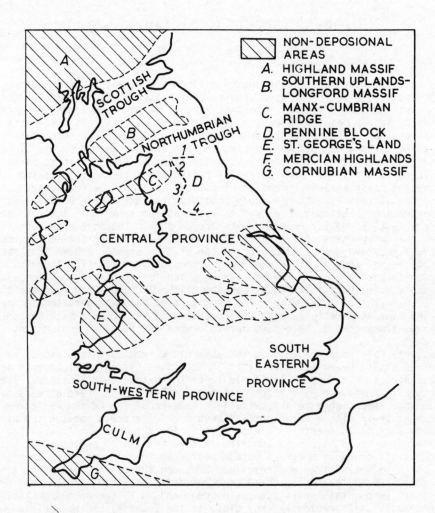

Fig. 38. Dinantian depositional areas (after George, 1969). Key: 1. Stublick faults; 2. Pennine Fault; 3. Dent Fault; 4. Craven faults; 5. Widmerpool Gulf.

## THE  DINANTIAN

Upper Devonian times in Brittany was a time of regression and restricted
areas of deposition.  "Domnonaea", the land area in the western English
Channel, was rejuvenated by the late Devonian earth movements and was to
spread molasse type debris into Breton areas by Tournaisian (Lower Dinantian)
times.  The effect of this important extensive uplift to the south of the
British area was to spread the sea northwards into the depressions on the
Old Red Sandstone Continent.  Attempts at this had already taken place in
Upper Devonian times (in Pembrokeshire and Breconshire, for example).  Now
the transgressions were to be more lasting.  The sea flooded into the Anglo-
Welsh Basin and into Southern Ireland.  Some partial invasion, at least,
of areas further east in Southern England is indicated by the occurrence
(underground) of Tournaisian limestone near Cambridge.  How far north the
Tournaisian transgression extended is difficult to assess.  It may have
reached into the "Central Province" (see fig. 38) of Lancashire, parts of
Derbyshire and Yorkshire, the Ravenstonedale area and the Irish Sea, but in
many parts of these areas the basal formations are not reached.  Further
north, evidence of marine formations older than Visean is virtually absent.
Continuity of deposition upwards from the Old Red Sandstone did occur, in
parts of Northumbria and especially in the Midland Valley of Scotland, but the
deposits are of non-marine character and even frequently red so that it is
difficult to assess the Devonian-Dinantian boundary.  In areas north of Central Ireland, too, Tournaisian rocks are lacking.  Visean ($C_2S_i$) ortho-
quartzites and conglomerates (Moy and Boyle Sandstones) form the Dinantian
base in the areas between the Curlew and Ox mountains in N.W. Ireland.
Fluvial and marginal marine detrital sediments were forming in discrete
basins of deposition fed by debris from the rising ancestral Ox-Curlew
uplands (Dixon, 1972).  In C.Scotland deposition in lower Dinantian times was
almost certainly in estuarine or lagoonal flats periodically flooded by
streams depositing cross-bedded lenticles and sheets of sand.  These cementstones are nearly 1300 m thick in the Lothian Basin (George, 1969).  Further
west in this Midland Valley trough, some cementstone type deposition was
followed by the eruption of the thick (1000 m) Clyde Lavas "whose accumulation in a great arc built a major landmass or archipelago that interrupted
sedimentation" (George, 1969).

There was marked differential subsidence over the British area in Dinantian
times.  Some areas subsided readily and in some cases very appreciably.  In
the Northumbrian trough (fig. 38) the Dinantian is almost 2200 m thick in
places.  In the Edinburgh district within the Scottish Trough the thickness
probably exceeds 2600 m.  In the Central Province the succession down to
lowest Visean levels is over 1800 m thick.  There may in places be a substantial thickness of Tournaisian beneath.  In the South Western Province,
thicknesses of 1200 m are reached in the Gower Peninsula and in the Mendips.
In the extreme south of Pembrokeshire that total increases to 1400 m.  On the
southern side of the Central Devon Synclinorium, the Lower Carboniferous
(almost equivalent to the "Lower Culm") is probably about 600 m thick.  On
the northern rim of that major downfold a thickness of about 750 m is
attained.  These Culm deposits of Cornubia occur again, but much more thickly
(2000 to 2600 m) in the extreme south of Ireland, especially in counties
Cork and Kerry.

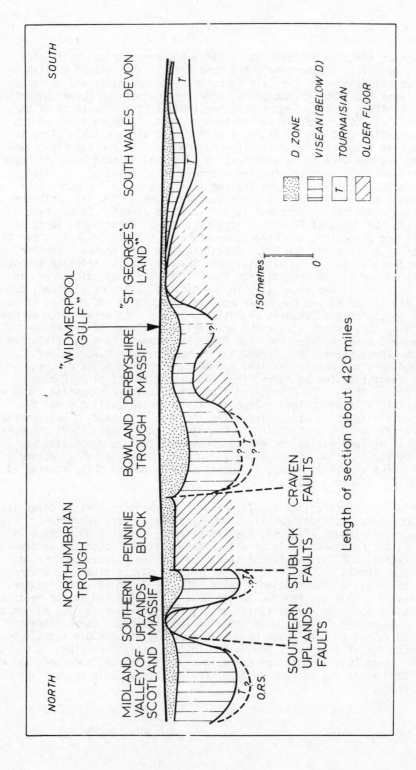

Fig. 39. Diagram section across the British Dinantian basins (after George, 1969). In vertical scale key, change 150 to 1500 metres.

Other areas however were persistently loath to subside even though some of
them never perhaps attained any worthwhile height.  They are shown in
fig. 38 and include St. George's Land, the Mercian Highlands (with extensions
northwards into Lincolnshire and Derbyshire), the Manx-Cumbrian ridge, the
Southern Uplands-Longford area and the Pennine Block, a fault-bounded area
that was not overwhelmed by deposition until well into Visean times ($S_2$).
Finally two marginal areas probably had considerable height and relief and
supplied debris to the neighbouring troughs - these were the Scottish Highlands in the north and a rising mass to the south of Cornubia (probably the
"Domnonaea" of Brittany and the Western Channel).  There is then considerable
lateral variation of thickness in the Dinantian when traced across the British
area, as shown in Professor George's section (fig. 39).  Note the thick and
the thin areas, the presence or absence of Tournaisian, how the greatest
lateral differences occur within the pre-D Visean unit and how these differences are beginning to even out in the topmost unit of the Dinantian.
Faults certainly play a part in controlling and determining the edges of the
"positive" (that is, loath to subside) areas.  Note, for example, the
remarkable correlation of isopachyte (thickness) lines in fig. 40 (again from
Professor George's work) with the bounding faults of the Alston and Askrigg
blocks.  The Southern Uplands Fault similarly controlled the southern side
of the Central Scottish Trough (fig. 42).  Over the Southern Uplands, the
Dinantian was either not deposited or only very thinly preserved, in areas
such as Thornhill and Sanquhar.  In parts of Ayrshire, the Dinantian is
barely 60 m thick.

Lateral variations occur also in successions when compared from basin to
basin.  In the Scottish Trough sandstones, shales and dolomitic limestones
(cementstones) pass up in the eastern areas into a rhythmic sequence of
sandstones, seat-earths, thin coals, shales and oil-shales.   These are
followed by the Lower Limestone Group, alternations of delta sands, paralic
muds and coals and clear-water marine calcarenites, "deposited in a basin of
pulsatory subsidence" (George, 1969).  These rhythmic repetitions form a
Yoredale type of sedimentation, prevalent also in various parts of the
Northumbrian succession and again over the Pennine Block (fig. 41).  Some of
the delta spreads, for example the Fell Sandstone, pushed out from a north-
easterly source.  Vulcanicity was intense in the Scottish Trough, with thick
accumulations in the Clyde Plateau area, the Garleton Hills and around
Arthur's Seat.  Extrusive activity in the Scottish Trough was greatest in the
Dinantian but persisted in places through into Namurian, Westphalian and (in
Ayrshire) even into Permian times (fig. 43).   The lavas and tuffs are all
part of an alkaline (sodic) magma series, as also is the suite of minor
intrusions associated with them.  By far the bulk of the Dinantian lavas are
olivine basalts.  Intermediate and acid members of the alkaline lava suite
include trachybasalts, trachytes, felsites and rhyolites.  These tend to be
found at higher levels of the lavas and are probably therefore a mature
stage in the volcanism.  Many necks and plugs of tuffs and agglomerates are
associated with the lavas.  It is tempting to see in this Carbo-Permian
igneous activity some early signs of crustal stretching, perhaps early
attempts to crack over a wide area from Rockall in the west to the North Sea
in the east.  Russell (1972) has hinted at an east-west acting tension over
Central Scotland initiating geofractures in late Tournaisian to Visean times.

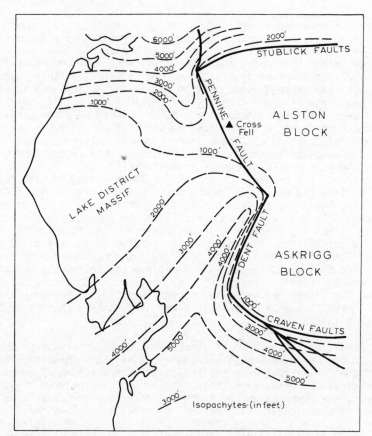

Fig. 40. Thickness variations of the Lower Carboniferous
in N.W. England (based on George, 1958).

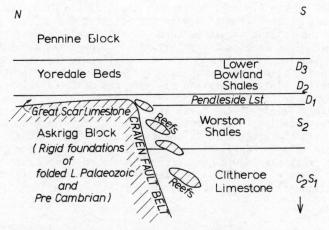

Fig. 41. General sketch section across the Craven
Fault zone.

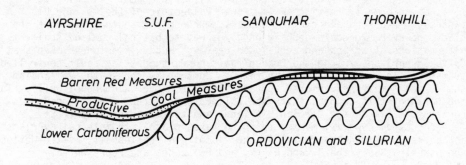

Fig. 42.  Section from Midland Valley to the South
Uplands (after Simpson and Richey, 1936).

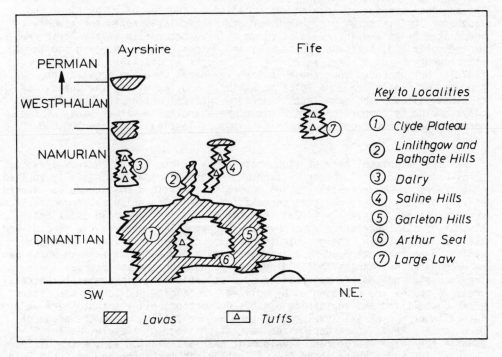

Fig. 43.  Time-space extent of extrusives in the Carboniferous of the Midland Valley (after Francis, 1965).

Lateral changes are particularly striking across the Craven faults, on the
southern edge of the Pennine Block (fig. 41). The Great Scar Limestone
($S_2$), unconformably resting on the Lower Palaeozoic of the Askrigg Block,
passes up into Yoredale cyclic measures. Beyond the southern edge of the
block, however, Yoredale deposition gives way to basinal Bowland sediments,
alternations of shales and cephalopod-bearing, muddy limestones. Reefs at
$C_2$ to $S_2$ horizons fringe the Craven fault-edge, suggesting fault-controlled
submarine slopes in lower Visean times. Further to the south, Bollandian
shales and limestones thicken appreciably within the "Widmerpool Gulf". To
the north of this trough, thick shelf limestones extend from $C_1$ to $D_2$
horizons. Towards the Mercian Highlands and the northern edge of St.
George's Land, however, marked thinning takes place with internal overlaps
and oversteps southwards until in the vicinity of Lilleshall, Little Wenlock
and Oswestry, only the $D_2$ and $D_3$ ($P_1$ and $P_2$) zones are represented.

On the southern side of this Midland-Welsh barrier, Tournaisian beds appear
in the Clee Hills and a fuller Dinantian succession carries on into the
Forest of Dean where sands appear in the higher Visean. These sandy incursions derived from a northerly source pushed on southwards with time into the
Bristol District and even into the north-eastern corner of the South Wales
Coalfield before the end of Dinantian times. Further south are the classic
carbonate sequences of Gower and the Bristol Avon. In Gower, muddy basal
sediments pass up through Tournaisian dolomites into oolites (Caninia
Oolite). These oolites have an upper erosive boundary marking emergence
before deposition of a calcite-mudstone "lagoonal" horizon. Higher in the
Visean occurs another (thicker) oolite formation - the Seminula Oolite. This
has an algal horizon at the top, also called a "lagoon-phase" by Dixon and
Vaughan. The $D_1$ and $D_2$ zones in Gower are particularly noted for pseudo-
brecciated units and interbeds of clays. These occur also on the North Crop
of the Coalfield, again in the same $D_1$ and $D_2$ positions. They are similar to
the clay wayboards of Derbyshire and Staffordshire, described by Walkden
(1972). He concluded that these English wayboards were the argillation
products of contemporaneous volcanic ashfalls and lay within the definition
of K-bentonites. The mammilated limestone surfaces and crusts beneath the
clays could, by comparison with Pleistocene examples in warmer parts of the
world, represent periods of subaerial exposure lasting 30,000 to 100,000
years.

It was in the Bristol Avon section that Vaughan first set up the now classic
coral-brachiopod zonal scheme. Dixon and Vaughan further successfully applied
that scheme to Gower. With its subsequent modifications the scheme has proved
of very great value in regional correlation and in Dinantian palaeogeographical
reconstruction. Recently (1973), Ramsbottom has introduced an additional
(lithological) correlative aid. He has suggested that between the Mendips and
the Southern Uplands, the pre-$D_1$ Dinantian succession comprised four major
cycles, each characterised by a transgressive and a regressive phase which can
be recognised by changes in lithology and fauna. On the shelf areas these
transgressive phases are represented by bioclastic limestones whilst the regressive phases are represented by oolites and calcite mudstones, the latter
often dolomitised and containing algae. Ramsbottom believes that there were
actual lowerings of sea level rather than a mere filling up of the areas of
deposition. Reefs start during the regression stage of a cycle, as for
example the thick Cracoe reef development near Grassington at the top of $S_2$.
The classic Lagoon Phase above the Caninia Oolite marks the top of Cycle 2 in
Gower whilst the Algal phase at the top of the Seminula Oolite marks the top
of Cycle 4. Within each major cycle a number of minor transgressions and

regressions occur when each major cycle is traced into the Northumberland
Trough where minor cyclicity occurs throughout.  Above the base of the $D_1$
zone this minor cyclicity is operative over the whole area.  Ramsbottom
divides the Dinantian into six major cycles, the fifth and sixth correspond-
ing to $D_1$ and $D_{2-3}$ respectively.

In Central Ireland the Dinantian is very extensive, with a thickness of about
1000 m, thinning towards the Leinster and Longford-Down massifs but swelling
to about 1300 m in the intervening Dublin Basin.  The most important element
of the Central Ireland succession is a great sheet reef covering a surface
area of over 7000 square kilometers.  In the south, towards Cork, the reef
is nearly 700 m thick with a mat of polyzoan and algal fronds, often in the
position of growth.  Contemporaneous volcanics occur in the Limerick Basin.
When traced across County Cork and the extreme south of County Kerry, the
carbonate successions give way to the "Carboniferous Slate", a thick pile of
cream, white and grey sandstones with dark to black slates.  The accumulation
appears to be the result of the rapid erosion of an uplifted folded area to
the south or south-west of Ireland.  It is interesting to note that geophysical results suggest a southward thinning of the Old Red Sandstone, in
the extreme south of Ireland, against a rising Lower Palaeozoic floor.

A similar southward change to Culm-type deposition must occur from Somerset
into Devon, but there is an awkward thirty mile gap, broken only by the
isolated exposure of Avonian-type Visean limestone at Cannington Park.  The
northern Culm near Bampton does include some coral-bearing limestones at
Westleigh but this is the southernmost limit for any possible application of
Vaughan's zones.  The bulk of the Dinantian on the north side of the Central
Devon downfold is formed by the Pilton Beds containing trilobites and
brachiopods that show an upward extension almost into the Pericyclus Zone.
Very thin alternations of shales, cherts and thin-bedded limestones follow
the Pilton Beds.  Pyritic black shales with siliceous partings take the
highest Dinantian succession well up into the Namurian.   These deposits
spanning the Lower-Upper Carboniferous boundary must represent very slow
deep-water deposition under poorly aerobic conditions.  They could indicate
almost the non-depositional environments of trenches along subduction zones.
There could be some substance in the case put forward by Johnson (1973) for
a closed plate suture somewhere across Cornubia marking the previous site
of a one-time Mid-European sea.  Johnson recalls the markedly differing
radiometric dates shown by Dodson and Rex to occur in belts across S.Cornubia
(a 365 to 345 m.y. southern belt followed successively northwards by belts
with 340 to 320 and 310 to 270 m.y. respectively).  Johnson points out that
thick spilites (with pillow lavas) and tuffs are widespread in the early
Culm of North Cornwall and are associated with radiolarian cherts and black
shales;  manganese ores have been worked in the lava.  Between Launceston
and Dartmoor the lower portion of the Dinantian consists of black shales
with quartzite layers;  the higher sequences are thin with shales and muddy
limestones, again representing quiet conditions.  As yet there was no sign of
a rising area somewhere to the south.  It may be that the narrow Dinantian
rim from Boscastle to W.Dartmoor was just too far north to record the northward spread of coarser debris from "Domnonaea".  In the succeeding Namurian,
however, the evidence was to be preserved.

## THE NAMURIAN

Sudetic (that is, end-Dinantian) earth movements, important on the continent, were not very effective in Britain, at least as far as folding and faulting are concerned though certain areas were uplifted causing marine regressions, sandy spreads into carbonate areas and local unconformities. There was one major effect however. A drastic change in sedimentation occurred, especially in the South-West Province and in the Central Province, with the "Millstone Grit" replacing the Carboniferous Limestone. The barriers across mid-Wales and Mercia were rejuvenated forming the Wales-Brabant Uplands (fig. 44). Their erosive products spread into the northern rim areas of the South Wales Coalfield, where quartz conglomerates and quartzites form the so-called "Basal Grits". Southward prolongations of the barrier may have spread down the lines of the Usk and Severn axes. Debris, in the form of quartz sand, spread also into some of the south-western rim areas of the Central Province, more especially in Denbighshire (Cefn-y-Fedw Sandstone) and Staffordshire. Rejuvenated uplifts persisted through Namurian times with deltaic spreads of felspathic sands coming southwards into the Central Province (there is a southward migration with time). The views of Gilligan and Sorby have been upheld recently that these "Millstone" grits were derived from Caledonian-deformed rocks of the northern continent to the north-east of England (Ramsbottom, 1969). Dating of mica from Marsdenian ($R_2$) grits in Derbyshire has confirmed the Caledonian age. Ramsbottom, in his valuable account of the Namurian of Britain (1969) shows how successively higher grits in the Central Province reached further and further south. In the early Namurian, the thickest grits are near Skipton; in $R_1$ times they are thickest near Kinderscout and in $R_2$ times near Macclesfield. Towards the top of the Namurian, more uniform sand spreads occurred over almost the whole basin. The Wales-Brabant uplands probably became much reduced in height by early $R_2$ times with muds replacing quartz sands in South Wales. Some deltaic spreads in the Central Province moved more rapidly down somewhat steeper slopes causing turbidite flows. Examples are the Mam Tor Sandstone and the quartzose grits in the grossly overdeepened Widmerpool Gulf in Derbyshire.

Oscillations of sea level were frequent in the Namurian and according to Ramsbottom some 60 marine bands, each with a diagnostic goniatite fauna, are recognisable in the British Namurian. A new goniatite fauna inhabited the area every 200,000 years. Ramsbottom calculates that the mudstones were deposited at about 1 ft in 4000 years. Within the fossil bands, benthonic faunas (brachiopods, crinoids, etc.) occur around the margins of the Central Province whilst goniatite-pectinoid faunas occupy the main central areas, which may have had relatively deep water. Muds with goniatite faunas are also characteristic of more open (arenite-free) waters in the Gower region of South Wales, where the Namurian is 700 m thick, nothing like the 2000 m coarser sequences of Lancashire.

North of a line more or less coinciding with the E-W position of the Craven faults, a change of facies occurs and Namurian areas to the north have a typical Yoredale cyclic facies of limestones, sandstones and shales with extensive coal deposits near the base in Scotland. Limestone horizons are traceable over wide areas as, for example, the Top Hosie (base of the Namurian) and the Index Limestone (top of the Limestone Coal Group) in the Midland Valley. Above the middle of $E_2$, the remaining Namurian (the Passage Group) is thin and sandy, with local breaks, many seatearths and occasional marine bands. There is however no big break within the Namurian as was

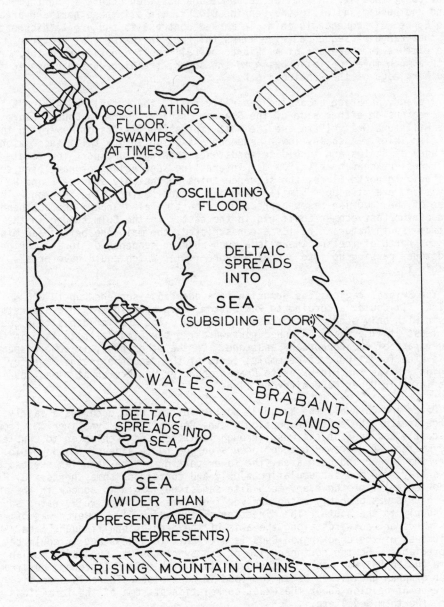

Fig. 44. Namurian Palaeogeography (after Ramsbottom, 1969).

previously thought. A similar sequence and position occurs in Northumberland and Durham and on to the Pennine Blocks. In all these northern areas, the $E_{2c}$ to $G_1$ sequence is thin, in marked contrast to the great thickness of that range of sequence (five times thicker) in the Central Province. The picture is therefore of a thinner overall shelf deposition in the north with a much more subsiding basin to the south, the hinge being close to the southern edge of the Pennine Block.

In Ireland, Namurian areas are generally small, the notable exception being the regions on either side of the Shannon Estuary. Here in County Clare, the Millstone Grit cliffs are the most impressive (Cliffs of Moher) in the British Isles. A shaley lower sequence passes up into a thick succession (R age) of shales and turbidite sandstones showing slumping. The total thickness reaches nearly 1500 m. Interesting sand volcanoes have been described. In north Clare, the sequence thins appreciably and the E and H divisions are missing. Similar thinning occurs south of the Shannon. The base of the Namurian rises appreciably with time also around the southern end of the Castlecomer Coalfield in the east. In the Culm trough in the extreme south Namurian relics are restricted, the main one being near Minane. In the north of Ireland there is a much thicker sequence of lower (marine) mudstones passing up into a Coal Measure facies which could have breaks within it.

In the Bristol region, the Namurian is a quartzose sequence up to 200 m thick. It could, according to Ramsbottom, have been formed in a discrete basin with only occasional access to the sea. The area of deposition could at least at times have extended northwards to the Clee Hills to take in the lower part of the Cornbrook Sandstone. On the other hand, in some respects this Clee formation even resembles parts of the sequence in the Oswestry region. Maybe the Mid-Wales to Brabant barrier was broken at times, as it was later in Upper Carboniferous times.

In Devon and Cornwall the Namurian portion of the Culm occupies a fairly large area of the southern portion of the Central Devon Synclinorium, from Boscastle and Bude on the west through Tavistock and Okehampton to Chudleigh and Exeter. A narrower outcrop occurs on the north side of the downfold. Over the larger southern area, the lower portion of the Namurian succession is shaley, the muds accumulating slowly and containing some cherts. By $H_1$ (Chokierian) times however, turbidite sandstones begin to appear in the areas. The debris was probably being derived from some rising or rejuvinated source of supply to the south. The direction of turbidity flow however was often westwards or eastwards along the axis of the depositional trough. The thickness of the Crackington Beds is difficult to assess but it could be appreciable. On the northern side of the Synclinorium, much of Namurian time (up to $R_2$) was one of muddy deposition with thin siltstones, but from then onwards greywackes began to flood in here also. There is some difference of opinion about the exact correlation of some of the formations here (Northam Beds, etc.).

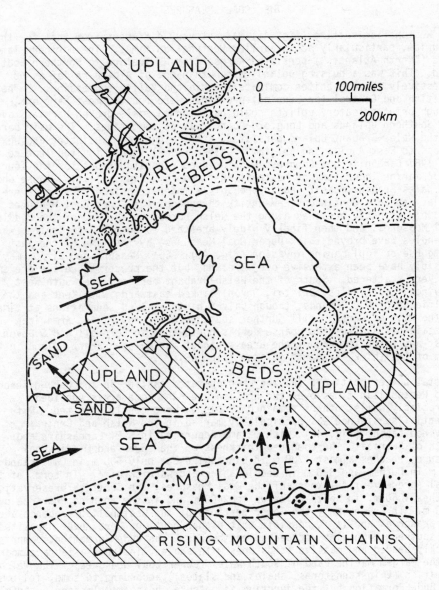

Fig. 45. Palaeogeography at time of the last Westphalian marine invasion (Cwmgorse or Top Marine Band) (after Calver, 1969).

## THE COAL MEASURES

The Westphalian setting (fig. 45) was not very different from that for the
Namurian, particularly so in earlier Westphalian times.  The northern landmass ("North Atlantis") persisted across from NW Ireland to Northern Scotland.  This was a pulsing upland rather than one of Alpine relief as no
excessively thick arenites dominate the Midland Valley successions.   The
Scottish and Central (or Pennine) provinces were for the most part one,
though brief pulsatory uplifts could have formed small broken barriers over
the Southern Uplands and Longford-Down areas at some times.  The main barrier
of the Wales-Brabant Upland persisted, stretching from E.Leinster and Wexford
across to East Anglia and Belgium.  Calver (1969) in his invited lecture to
the 1967 Carboniferous Congress has included reconstructions for the four
major marine incursions of Westphalian times (his figs. 11-14).  These show
the Wales-Brabant massif (again probably a moderate upland with no severe
relief) to be a continuous E-W entity throughout the Lower Coal Measures,
possibly slightly breached along the Welsh Border by the end of the Middle
Coal Measures, and then finally widely breached to beyond Oxfordshire (where
boreholes have proved thick Upper Coal Measures) by the time of the last
(Cwmgorse or Top) marine invasion (fig. 45).  Coal Measure deposition may
possibly have been extensive over Ireland, but the preserved relics are small
and very scattered.  South of the Welsh-Brabant barrier, the South-West
Province (South Wales, Bristol, Devon) linked eastwards with Kent and the
coalfields of N.E. France, though Calver hints at land separations at times
to the west of Kent.  Still further south were the now rising Armorican
mountain belts, flushing coarse debris northwards to Cornwall and S.Devon at
first and later to much of the area south of the Wales-Brabant Upland in the
form of the thick "Pennant" molasse (fig. 45).

Crustal pulses continued (as they had operated earlier in the Carboniferous)
into Westphalian times over the British area and some areas subsided
appreciably whilst others remained stable or were gently uplifted.  Differential subsidence was particularly marked in the Welsh and Central
Province basins (fig. 46), with maxima occurring in S.E. Lancashire and
north of Swansea.  The combined thickness of the Lower and Middle Coal
Measures reaches almost 1700 m in Lancashire but only 500 m in Cumberland,
650 m in Durham and 320 m in Denbighshire and Warwickshire.  In terms of the
total (preserved) thickness of the Coal Measures, the Scottish preservation
reaches 1060 m, in South Wales up to 2400 m and in the Pennine Province over
3000 m.  In Devon, only the lower portions of the Westphalian sequence are
now preserved and it is possible that higher portions were never deposited as
the southern mountains encroached northwards.  The Bude Formation, comprising
thickly-bedded and massive sandstones with shales, includes a fauna comparable
to the Margam Marine Band of South Wales (Lower Coal Measures).  The Welcombe
Formation (thin sandstones, shales and slates), according to some, follows
the Bude Formation but the junction is obscure, being complicated by infold-
ing and inversion.  The Welcombe Beds could even be equivalent to high
horizons in the Crackington Formation.  Other formations in the Barnstaple-
Bideford area, such as the Cockington Beds, Abbotsham Beds and Instow Beds,
are believed by Reading and others (1965) to be stratigraphical equivalents
of early Westphalian age.  Reading believes that these different stratigraphical sequences at about the Namurian-Westphalian boundary have been
brought into juxtaposition in these N.W. Devon coastal exposures by thrusts.
As a result southward-transported sandstone-shale sequences of paralic

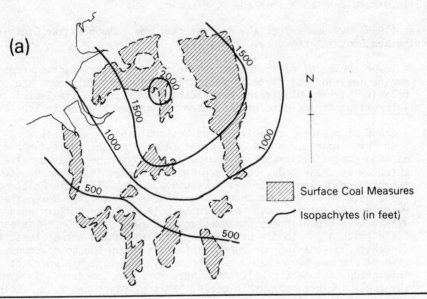

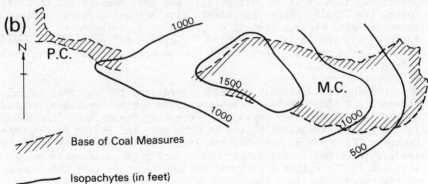

Fig. 46.  Thickness patterns of parts of the Coal
Measure successions in (a) Pennine and Midland Coalfields
(modiolaris and lower similis-pulchra zones) and (b)
South Wales (Middle Coal Measures).  (a) after Trueman,
1947;  (b) after Owen, 1964.  M.C. Main Coalfield;
P.C. Pembrokeshire Coalfield

facies now lie south of northward-transported turbidite sequences (the wrong way round, as it were). Apart from the problematical Ugbrooke Park Beds near Newton Abbot (they could also be of Crackington age) no higher Westphalian strata are known from the Devon Culm.

Calver (1969) has summarised the general features common to the South-West, Pennine and Scottish provinces. They are:

(1)   A lower series of marine bands, up to ten in number, marking frequent marine invasions at first in Westphalian times. Coal seams are relatively thin, plants and musselbands are sparse. The Gastrioceras subcrenatum Marine Band has not been recognised however in Scotland.

(2)   A group of measures straddling the Westphalian A - Westphalian B boundary. Apart from the Clay Cross (Amman in South Wales) Marine Band marking that stratigraphical boundary, truly marine invasions are absent. In these beds, coal development reaches its maximum with numerous thick and economically valuable coals. In South Wales many seams between 4 and 9 ft thick occur and in the Vale of Neath there is one seam 18 ft thick. The best anthracite coals in Britain occur within this group in the northwestern quadrant of the South Wales Coalfield (fig. 47). In the Lancashire Coalfield there are at least forty distinct coal-bearing horizons within this group though of course several are not of workable thickness. In the East Midlands, coals in this group include the Silkstone, Deep Hard and Top Hard. In Ayrshire the coals include Pennyvenie coals, Camlarg, Major and Main. Calver points out that it is in this part of the Coal Measure sequence that the non-marine lamellibranchs and plants attain their greatest variety and abundance with various species of Carbonicola, Anthracosia, Niadites, Anthraconaia and Anthracosphaerium. Several of these genera were to die out or become much less common in the next group of strata (see Calver's fig. 4).

(3)   An overlying series of marine bands straddling the Westphalian B-C boundary. In South Wales there were seven marine invasions, including the well-known and very persistent Cefn Coed Marine Band (the Mansfield in the East Midlands, Skipsey's in Scotland) and ending at the Upper Cwmgorse (Top) marine horizon, the last invasion of the British area by the sea in Carboniferous times. This last horizon is again not known in Scotland. The return of the marine incursions in this third group of strata corresponds to a reduction in coal formation. In Scotland and along the northern edge of the Wales-Brabant barrier, the "red-bed" facies is developed locally in the uppermost beds.

In the remaining sequences (that is, in the Upper Coal Measures), the general similarity which had existed so far is no longer recognizable between the three areas. South of the Wales-Brabant uplands, coal-forming (paralic) conditions persisted in South Wales, the Forest of Dean and in the Bristol-Somerset Coalfield, but there were frequent periodic influxes of thick ill-sorted, subgreywacke-type Pennant sandstones. Kelling (1974) has reconstructed the sedimentological pattern of the Lower Pennant Measures (upper portion of Westphalian C) in South Wales (fig. 48). Rivers were flowing northwards into South Wales spreading fluviatile sands and muds. The presence of lagoonal-type and bay sands and also barrier bar sands in the north-eastern quadrant of the Coalfield and again in the north of the Forest of Dean

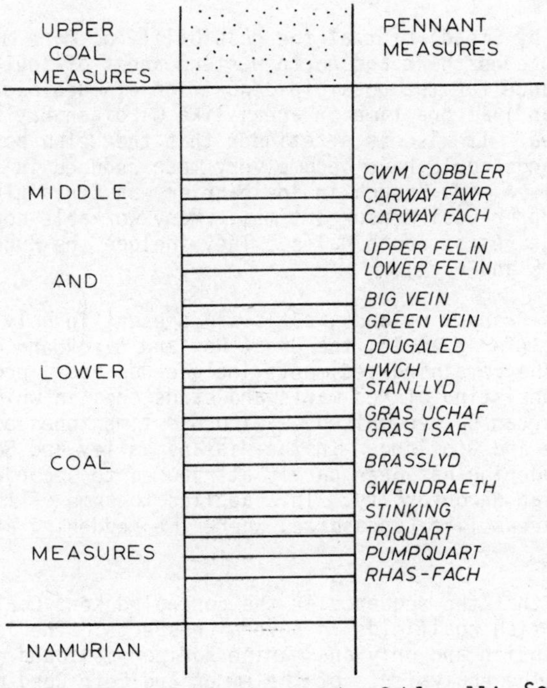

UPPER
COAL
MEASURES
PENNANT
MEASURES
MIDDLE
AND
LOWER
COAL
MEASURES
CWM COBBLER
CARWAY FAWR
CARWAY FACH
UPPER FELIN
LOWER FELIN
BIG VEIN
GREEN VEIN
DDUGALED
HWCH
STANLLYD
GRAS UCHAF
GRAS ISAF
BRASSLYD
GWENDRAETH
STINKING
TRIQUART
PUMPQUART
RHAS-FACH
NAMURIAN

Fig. 47. Coal seams in the area north of Llanelli, South Wales.

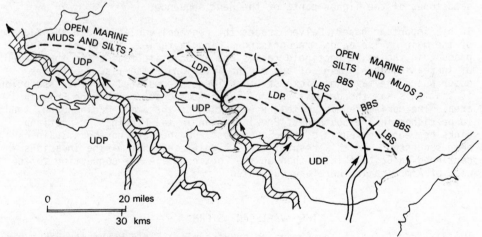

OPEN MARINE
MUDS AND SILTS ?
UDP
LDP
OPEN MARINE
SILTS AND MUDS ?
LBS
BBS
LDP
UDP
BBS
BBS
LBS
UDP
UDP
0
20 miles
30 kms
UDP Upper delta plain sediments
LDP Lower delta plain sediments
BBS Barrier Bar sands
LBS Lagoonal and Bay sands
Suggested rivers

Fig. 48. Lower Pennant Measures sedimentation in South Wales (after Kelling, 1974, fig. 48).

(studied recently by Stead) hint at the possibility of more marine conditions towards more northern and north-western areas of South Wales.  This
has some significance for the possible deposition of much higher Carboniferous (Stephanian?) at one time in areas like Cardigan Bay and parts of
the South Irish Sea.  It also is a reminder that the Welsh portion of the
Wales-Brabant uplands could have become very much reduced in size by late
Westphalian times.  A wide breach in the barrier was certainly created by
the beginning of Upper Coal Measures times.  Many workable coals occur in
the Upper Coal Measures of South Wales.  They include the Hughes and the
Swansea seams (4, 5 and 6 ft).

North of the Wales-Brabant uplands, coals are present in only a few areas or
at limited periods (for example, the Great Row and Blackband groups of North
Staffordshire).  The remaining sediments include red-beds, probably of
primary origin, consisting of red marls and sandstones in which large-scale
rhythmic sequences can be recognised.  Spirorbis limestones occur in the
Midlands, Pennines and Scotland.  In the Midland Valley and Southern Uplands
coalfields the reddening has been partly attributed to secondary oxidation
from the sub-Permian unconformity.  This applies to some of the reddening
patterns also in areas like Lancashire, where the reddening affects varying
horizons.

Calver points out that the sequence in the concealed Kent Coalfield differs
from the other British coalfields in several respects.  The lower Westphalian
there lies on Dinantian and only one marine horizon is found in these
lowest measures.  The equivalents of the Amman and Cefn Coed marine bands
are however strongly developed.  An unconformity lies beneath the Pennant
sandstones of the higher parts of the Kent sequence.

In his important paper, Calver traces the regional variations in the faunas
of the main marine bands, recognising a Lingula facies, a Myalina facies, a
productoid facies, a pectinoid facies and a goniatite facies (in that order
of increasing development of any marine invasion).  The last two facies
occur where the maximum thickness of sediment accumulated and the assumption
is that this was the deepest part of the contemporary sea of the British
area.  The marine faunas, with the exception of the Amman Marine Band, tend
to be richer in the South-West Province than in the Pennines or Scotland.
Links between the faunas of the South-West Province and northern Spain have
been suggested.  In this respect it is significant that marine invasions
continued in Britain later than on the Continent.  A sea connection to the
west of Armorica is therefore suggested.


## THE  VARISCAN  STORM

The Variscan (also called Armorican or Hercynian) Earth Movements had their
greatest effects in Britain in late Carboniferous times, mostly as late as
the Stephanian.  The post-Variscan unconformity is displayed in many areas.
Fig. 49 shows how the Permian rests on folded units of the Carboniferous
between Durham and Yorkshire.  That Permian however is Upper Permian.   In
E.Devon much lower Permian rocks rest unconformably on Devonian or Culm.
In many other areas the first deposits to occur unconformably on the eroded
Variscan folds are Triassic (often higher Triassic beds).  Good samples of
the sub-Triassic unconformity are to be seen at Sully  and Barry, in the Vale
of Glamorgan, and again at Clevedon.  There have in the past been many

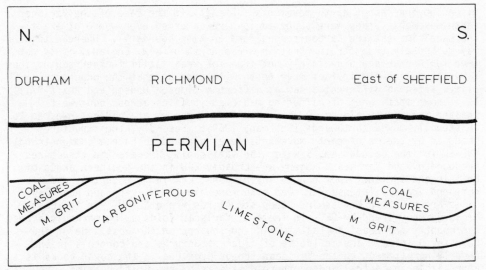

Fig. 49.  The unconformity beneath the Permian in N.E.
England.

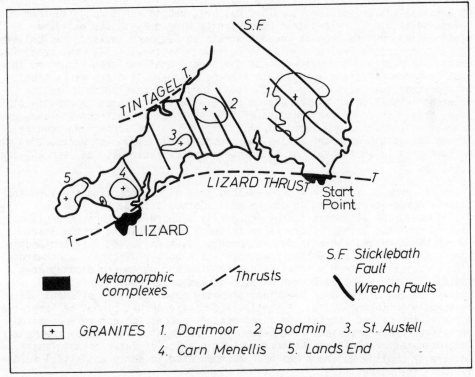

Fig. 50.  Principal faults and granites of S.W. England
(based on Hosking, 1962).

misconceptions about earth movements.  Firstly in the case of the Variscan
earth movements, they were supposed to have occurred everywhere at exactly
the same time (at the Carboniferous-Permian time boundary).  Whereas this
may be broadly true for the British areas north of S.W. England, it is not
true for Cornubia and certainly not true for areas still further south. The
Variscan pulses did in fact move north with time so that the orogenic
climax affected different areas at different times.  Dodson and Rex (1971)
have shown that zones of differing radiometric dates across southwest
England range over almost 100 m.y., while the phases of folding appeared to
be steadily moved northwards (Dearman, 1969).  Secondly, the structures
produced by the major earth movements were supposed to have a common trend
(NE-SW for the Caledonian, E-W for the Variscan).  Whereas the structures
produced by the Variscan Orogeny are broadly E-W in the southern areas of
Britain and Ireland, they can be anything but E-W in more central and
northern areas.  Intense Variscan structures in the Malvern and Abberley
hills are N-S.  In Lancashire steep structures trend NE-SW.  In the East
Midlands they trend NW-SE.  In Ireland, Variscan folds swing markedly across
the country from west to east turning northwards on the west side of the
Leinster Granite.  The influence of older structures and controls in the
underlying basement on the Variscan trends produced in the cover rocks is
appreciable in central and northern districts of the British Isles.  Several
Caledonian and pre-Caledonian faults moved again in response to the Variscan
stresses.

A third misconception is that there was a marked violent change of structure
at a so-called Variscan Front.  This was believed to run through southernmost
Wales and across S.Ireland from Dungarvan to Dingle Bay.  Its eastward
continuation, buried beneath younger strata in Southern England, was believed
to reappear as the Faille du Midi in Belgium.  Matthews (1974) has recently
critically reviewed the evidence for the traditional Variscan Front and shows
that it means different things in different places.  It often marks a belt
of important thickness changes with local thrusting developing where the
basement is shallowing as in Belgium and S.W. Wales.  Facies changes in other
areas result in consequent changes of fold-style as in Southern Ireland,
thereby accentuating the supposed continuity of a line across the country.
If the Variscan Front runs across the southern peninsulas of Wales, then it
certainly fails in the Vale of Glamorgan where Variscan dips of over 25° are
a rarity.

There may even be a fourth misconception, this time concerning the supposed
Lizard-Dodman-Start Thrust in southernmost Cornubia (fig. 50).  Sadler
(1974) has drawn attention to this generally assumed structure, and claims
that it does not exist.  There is no trace of overthrusting at the Start
Schist boundary and there is no metamorphic basement in the Roseland-Dodman
area.  According to Sadler the Start Schists probably disappear northwards
down successively steep northward-downthrowing and northward-dipping step
faults.  In the Roseland-Dodman area, a series of thrusts (carrying
Ordovician quartzites over Devonian) probably resulted by decollement off a
basement high to the south.  A single large-scale thrust cannot be identified
north of the Lizard Complex and a sedimentary succession can be recognized
in alleged crush breccias.  The metamorphic basement, overlain by the Menaver
Conglomerates are left at the surface by a complex of later normal faults.
An already shallow basement has been accommodated in early anticlinal folds
by local thrust-faulting.

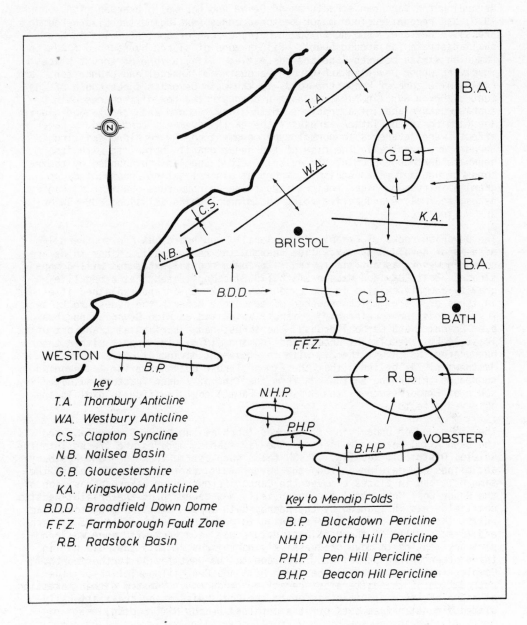

key

T.A.    Thornbury Anticline
W.A.    Westbury Anticline
C.S.    Clapton Syncline
N.B.    Nailsea Basin
G.B.    Gloucestershire
K.A.    Kingswood Anticline
B.D.D.  Broadfield Down Dome
F.F.Z.  Farmborough Fault Zone
R.B.    Radstock Basin

Key to Mendip Folds

B.P.    Blackdown Pericline
N.H.P.  North Hill Pericline
P.H.P.  Pen Hill Pericline
B.H.P.  Beacon Hill Pericline

Fig. 51.   Pre-Mesozoic structure of the Bristol district
(after Moore, 1951).   B. A. Bath Axis; C. B. Central Basin.

Regarding the Variscan structures of Devon and Cornwall, Dearman (1969, 1970) has recognised four major tectonic zones (excluding the Lizard-Dodman areas).  They are, from north-south:  (i) a zone of upright folds in West-phalian strata (as around Bude);  (ii) a zone of sliced overturned folds in Namurian strata between Bude and Boscastle;  (iii) a zone of thrust slices involving Upper Devonian and Carboniferous near Tintagel and Launceston;  and (iv) a wide zone of recumbent folds in the main Devonian tract south of the Central Devon Synclinorium.   Dearman demonstrated how differences with suprastructure and infrastructure (with slaty, second and crenulation cleav-age forming in the latter) arose as the earth movements continued.   Later effects were reversals to normal type movement northward along existing thrusts and eventually the rise of the great granite batholith with its separate Dartmoor, Bodmin, St. Austell (etc.) cupolas, surrounded by their zones of contact metamorphism.  Important mineralisation accompanied the granite intrusion.  Near the granites there are numerous granitic "elvan" dykes, tourmaline and aplite veins and mineral veins all along ENE-WSW or NE-SW lines.

The Devonian rocks of Exmoor dip regionally southwards as the southern limb of a major upfold whose axis lies near to the north coast.  Minor folds are common however, as a visit to the Ilfracombe coastal exposures will demon-strate.  Bott, Day and Masson-Smith (1958) demonstrated that a negative gravity anomaly existed in the Exmoor-Quantock region and explained it either in terms of a 4 km thick sequence of sandstone beneath the oldest rocks seen (Lynton Beds) or (preferably) a thrust which had carried Devonian northwards over Devonian and Carboniferous.  The thrust had a horizontal transport of at least 14 km.  Results by Brooks and Thomson (1973) can support either theory but prefer the thrust theory, with the movement having exceeded 25 km. Matthews (1974) believes that the Exmoor Thrust does not exist.   Devonian sequences thickening northwards from the line of a deep fracture situated south of Exmoor (such as the Brushford Fault) could suffice to explain the observed effects.

The influence of underlying structures (such as the Lower Severn Axis and the Bath Axis) becomes evident as the Variscan structures are traced into the Bristol District (fig. 51).  NE-SW folds such as the Thornbury and Westbury anticlines are developed along the Severn Axis. Earlier movements along the same belt had in places removed the Carboniferous cover before deposition of the Upper Coal Measures.  The essentially N-S downwarp of the Bristol-Radstock coalfields was influenced by the nearby Bath axis.  This warping took place whilst the Mendip upfold was beginning to develop.  As a result the eastern end (Beacon Hill) of the Mendip structure was held back with more western portions wanting to move progressively northwards.  This they did but in three other broken periclines, each one to the west moving further north. Complex fractures separate the North Hill and Pen Hill periclines.  The retardation of the easternmost pericline produced at Vobster either thrusting of nappe outliers of Dinantian on to the Coal Measures or gravitationally-glided Dinantian fragments off the upfolded Beacon Hill region.

The Variscan structures in Wales are shown in fig. 52.   Earlier Caledonian structures such as the Bala Fault and the Teifi and Towy anticlines were rejuvenated.   Eastward splays of the Bala fracture suffered transcurrent movements including rotation of the Minera-Llangollen region and producing rotary faults (Aqueduct Fault, etc.).  Upwarping along NNW-SSE lines along

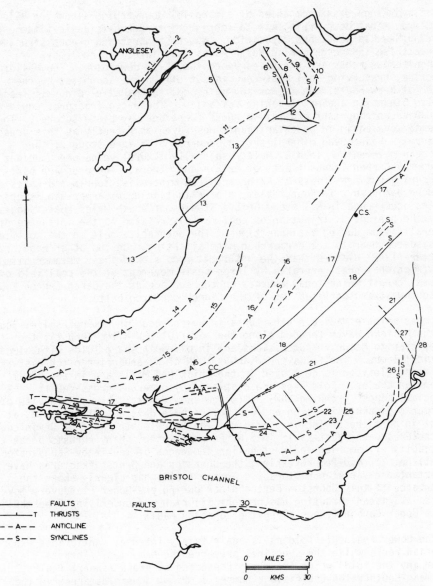

Variscan structures in Wales. Key: 1 Malldraeth Syncline; 2 Berw Fault; 3 Dinorwic Fault; 4 Llandudno Syncline; 5 Conway Fault; 6 Vale of Clwyd Syncline; 7 Clwyd Fault; 8, 9 and 10 Inner Crescentic Anticline, Outer Crescentic Anticline, etc.; 11 Derwen Fold. 12 Bryneglwys Fault, Llanelidan Fault, Aqueduct Fault, etc.; 13 Bala Fault; 14 Teifi Anticline; 15 Central Wales Syncline; 16 Towy Anticline; 17 Church Stretton-Careg Cennen Disturbance; 18 Swansea Valley Disturbance; 19 Benton Fault; 20 Ritec Fault; 21 Vale of Neath Disturbance; 22 Pontypridd Anticline; 23 Caerphilly Syncline; 24 Moel Gilau Fault; 25 Usk Anticline; 26 Forest of Dean Syncline; 27 Mayhill-Woolhope Anticline; 28 Malvern Fault; 29 Folds in South Pembrokeshire; 30 Folds in Gower.

Fig. 52. Variscan structures in Wales (Owen, 1974).

the Clwyd Range was accompanied by a parallel downwarping along the Vale
of Clwyd.  That downwarping was to continue into Permo-Triassic times preserving New Red Sandstone deposition.  Numerous horst and graben structures
characterise the neighbouring North Wales Coalfield.  In Anglesey,
Carboniferous rocks are preserved along two NE-SW downfolds.  The Malldraeth
Syncline, preserving Coal Measures, is cut off to the southeast by the
powerful Berw Fault.  On the nearby mainland, the parallel Dinorwic fracture
faults Dinantian against Arvonian volcanics.  These two fractures could well
lie above an important crustal weakness (dare one say plate suture).  They
were to move again in post-Variscan times (even as recently as this present
century).  Widespread mineralisation occurred at a late stage of the
Variscan movements.  In N.E. Wales especially mineral veins are abundant
along the Carboniferous tracts of Halkyn and Minera.  The ores are post-
fracturing but pre-Triassic.  Zinc and lead mineralisation in mid-Wales could
also be Variscan.  In South Wales, mineralisation near Carmarthen is updated
by its occurrence in Millstone Grit.  Studies of South Wales coals suggest
that the present distribution of coal rank is related to the influence of
mineralisation during Variscan times.  The anthracite belt in the coalfield
is situated near to the Carmarthen mineralisation.  On the other hand,
Trotter (1948) claimed that the anthracite coals owed their metamorphism to
the frictional heat generated by large scale movement of the coalfield over
a basal thrust which today emerges on the surface as the Careg Cennen Fault
(a southwestward continuation of the Church Stretton belt).

The most intense Variscan structures in Wales occur in Pembrokeshire. South
of the Milford Haven (eroded along the powerful E-W Ritec Thrust), the
Ordovician to Namurian successions are involved in sharp folds with almost
vertical limbs.  The folds trend about $10^{o}N$ of W, and often pitch eastwards.
Incipient cleavage is developed in the thick Red Marls and in the basal
Dinantian shales.  Fold axes are displaced by numerous cross-faults,
especially those trending NNW-SSE (which show dextral shear).  In view of
Dearman's findings in Devon and Cornwall, these dextral shifts could, at
least in part, be of Tertiary origin.  The Sticklepath Fault of Devon could
be the Freshwater West Fault in S.W. Pembrokeshire.  Many thrusts break the
continuity of Namurian and Westphalian sequences of the Tenby-Saundersfoot
coastline.  In mid-Pembrokeshire, the Johnston and Benton fractures have had
important pre-Variscan histories.  George (1963) has clearly shown the
importance of this Johnston-Benton block during pre-Upper Llandovery movements.  Variscan thrusting has brought Silurian and underlying Precambrian
on to Upper Carboniferous along the Johnston Fault.

In the Gower Peninsula folding is again fairly intense.  Anticlines expose
Devonian rocks while the synclines preserve the Namurian.  Thrusts
accompany the folds which are much transected by cross (shear) faults.  In
the main South Wales Coalfield a number of minor downfolds preserve the
highest Coal Measures of that area.  Important east-west faults include the
complex Moel Gilau Fault (near Port Talbot).  Two important wrench faults
dominate the Neath and Tawe valleys.  These NE-SW fractures are accompanied
by sharp parallel folds.  The fractures seem to extend northeastwards into
the Welsh Borderland, the Tawe system into Titterstone Clee and the Neath
Fault into the Abberley Hills.  A dyke occurs along the Neath Belt at
Bartistree (near Hereford).  The Tawe and Neath belts of disturbance must
lie above old fractures in the underlying basement (which may not be all that
far beneath the Devonian cover).

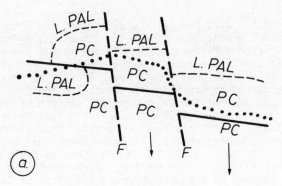

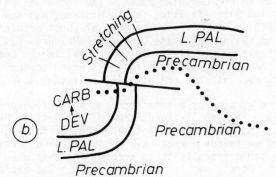

Fig. 53. Two versions of the origin of the Malverns structure, (a) after Raw; (b) after Butcher.

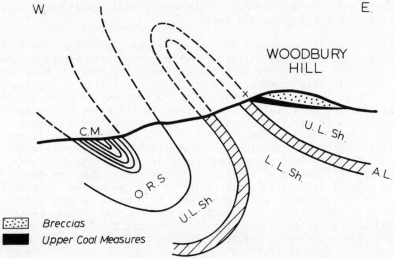

Fig. 54. Section across the Abberley Hills (after Mykura).
A.L. Aymestry Limestone, L.L.Sh. Lower Ludlow Shales,
U.L.Sh. Upper Ludlow Shales.

One of the most intense Variscan belts occurs along the Abberley-Malvern
line.  The northern (Abberley) portion is really a less eroded suprastructure
of the more deeply-eroded Malverns core.  In the latter Precambrian gneisses
break through covers of Palaeozoics (on the west) and Triassic (on the east)
pointing to a complex history of fracturing.  Brooks has shown that important
early movements occurred before Silurian deposition began.  Many theories
have been evolved to account for the geological pattern of the Malverns.  Two
of these ideas are shown in fig. 53.  Raw believed that the exposed Precambrian was part of a giant thrust nappe, dropped down along important
boundary fractures.  Butcher envisaged a major monoclinal structure with only
local thrusting on its vertical limb.  In the Abberley Hills westward over-
folding affects Silurian, Devonian and Upper Carboniferous rocks.  This E-W
compression must however have occurred before the close of Coal Measure
deposition because high Coal Measures seem to lie flat across the folds on
Woodbury Hill (fig. 54).  Breccias succeed the unaffected high Coal Measures.
In the Trimpley Inlier further north, small patches of Middle Coal Measures
appear to lie across folded Devonian rocks.  The intra-Westphalian and
Stephanian folding and block-faulting in the Midlands has been unravelled in
detail by Wills (1948 and 1956).  Two periods of considerable intensity
occurred, one at the beginning of the Upper Coal Measures (as for example
the famous Symon "Fault", actually an unconformity, in the Coalbrookdale
Coalfield), the second in high Stephanian times prior to the deposition of
the Clent-Haffield breccias (the end of Wills' palstage 4b).

N-S aligned Variscan structures extend from the Bristol-Severn area through
the Malverns and Abberleys to Staffordshire and the Western Pennines.
Folding along this line was often intense and any overfolding or thrusting
is usually directed westwards.  Local E-W compression was active over a long
period of time and the difficulty is to account for this when the overall
regional compressive stress was in a N-S direction.  One explanation involves
the inward movement of the St. George's and East Anglian blocks, producing
intervening E-W compression.  Another possibility could involve early
(abortive) attempts at E-W stretching in areas such as the North Sea.  Tension there and in areas further to the west ( Irish Sea?) could result in E-W
compression in the intervening areas.

In Northern England movements took place in the vicinity of the Alston-
Askrigg blocks.  Studies by Trotter and Wager indicate wrench movements along
bounding faults such as the Stublick and Craven systems.  These lateral movements appear to have produced an anticlockwise rotation of the (combined)
blocks with at the same time some sharp E-W compression along the Dent line
(adjacent to the Dent Fault).  Vertical movements along the western boundary
fractures were if anything opposite to those which they were to suffer in
later (Tertiary?) adjustments.  Similarly the Pennine upfold is really a
late (again Tertiary) structure as George (1963) has shown.  In the Lake
District, Variscan folding was comparatively gentle, the area being domed.
The outward inclination of the Carboniferous strata does not generally exceed
$30^{o}$ and some of this tilt is due to Tertiary accentuation.  The Carboniferous
rim is cut by very many faults which must have been, to some extent,
initiated by the Variscan movements.  Renewed movements have certainly
occurred however as so many of the fractures cut the Permo-Triassic rim.
Mineralisation, perhaps connected with some revival of granite basement,
occurred along several fractures.

In the Midland Valley of Scotland there are two main sets of Variscan
structures, those with a "normal" E-W trend and those with a north-east
(or even north-north-east) trend.  The interlocking nature of the two sets
is seen in the complex structural depression of the Central Coalfield.  East
of this basin caledonoid structures predominate;  to the west, E-W structures
are most common.  Important E-W fractures include the Campsie Fault with a
southerly downthrow reaching 1000 m.  The Ochil Fault, to the north, has an
even greater southerly downthrow (up to 3300 m).  The Highland Boundary and
Southern Uplands fractures behaved as sinistral transcurrent faults during
the Variscan deformation.  It is almost certain that the Great Glen Fault
also moved at this time but it is not easy to prove Variscan shift.

To understand the effects of the Variscan Orogeny in North and Central
Ireland one has first to bear in mind a number of contributing factors:
(i) the form of the sub-Devonian floor;  (ii) the presence of hard cores
such as granites, thick volcanic sheets, even thicknesses of tough sedimentary formations both within the basement and the overlying younger blanket;
(iii) the trend and form of these bulwarks, whether they be in the basement
(or as buried topographies on its surface) or in the blanket;  (iv) the
variation in thickness of the blanket, more especially in the direction of
major compression.  In the case of Central Ireland the main internal
obstacles to the S-N Variscan compression of the Devonian and Carboniferous
sedimentary sheets was undoubtedly the great Leinster Massif with its hard
cores of granites, diorites and Bala volcanics.  Other contributing influences
were the Connemara mass in the north-west, the Longford-Down Massif in the
north-east, the caledonoid trend of the Caledonoid structures in the Lower
Palaeozoics, and the way in which the Old-Red Sandstone wedges out northwards
from its great thicknesses (3000 to 5500 m) in the south.  Further hidden
influences have been also suspected.  Capewell (1957) believed that the broad
arch of the Comeraghs (surrounded by more severe folding) was due to the
presence of a deeper, possibly plutonic, mass as yet unexposed.  A similar
granite core could occur in the Slievenaman Inlier to the north.  The most
outstanding Variscan feature in Central Ireland is the way in which the fold
axes turn north-eastwards or even north-north-eastwards towards the western
side of the Leinster Massif.  Traced through to the Dublin-Loughshinny coast,
the folds then gradually turn back to an easterly direction.

The Variscan earth movements impressed on Southern Ireland its great structural (and topographical) features, the Devonian and Carboniferous rocks
being thrown into a large number of folds which curve from an E-W trend in
the Cork-Youghal area to ENE-WSW in the southwestern headlands and bays.
Complex upfolds include the Mangerton, Great Island, Clashmore, Ballycolton,
Clonakilty and Kilcrohane anticlines.  The synclinoria include those of
Bantry Bay, Blarney, Ardmore, Cork, Cloyne and Minane.  The axes of these
major folds rise and fall as they are crossed by culminations and depressions
with thereby westward and eastwards plunges giving frequent boat-shaped outcrops of the synclinal Carboniferous rocks.  The folding in southernmost
Ireland is of the cleavage or parallel type as opposed to the concentric
type in the "ground swell" area further north.  There has been overturning
to the north.  The shales and silstones display strong axial plane cleavage
whilst in many areas the siltstones show fracture cleavage.  Superimposed on
all the major folds are a very large number of minor flexures, ranging from
hundreds of feet to mere inches in amplitude, especially in the argillaceous
formations.  Complex fault patterns include ENE-WSW thrusts, powerful wrench-
faults (often with dextral shift) trending WNW-ESE in the Sneem area and a

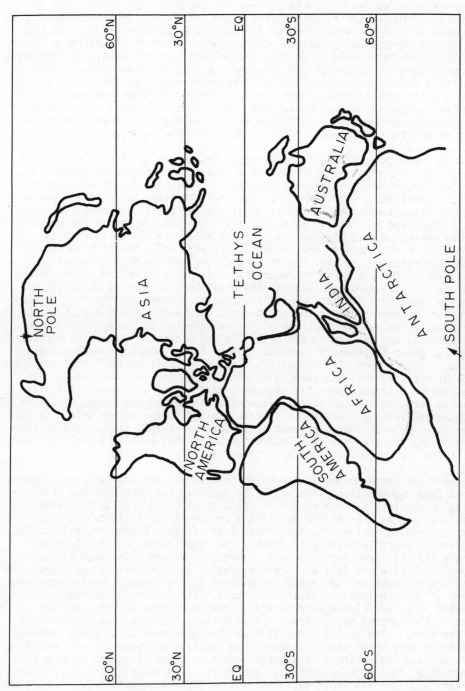

Fig. 55. The Permo-Triassic World (from The Breakup of Pangaea by R. S. Dietz and J. C. Holden Copyright ©1970 by Scientific American, Inc.

large number of cross-fractures ranging from NW-SE through to NNE-SSW.  The
thrusts are particularly concentrated along a line running from south of
Killarney to Millstreet and Mallow (the "Variscan Front").


## Suggested Further Reading

Anderson, J.G.C. and Owen, T.R. 1968.  The Structure of the British Isles.
    Pergamon Press, Oxford.
Calver, M.A. 1969.  Westphalian Britain.  Compt. Rend. 6me. Congr. Strat.
    Carb. 1, 233.
Dearman, W.R. 1970.  Some aspects of the tectonic evolution of southwest
    England.  Proc. Geol. Ass. Lond. 81, 483.
Dodson, M.H. and Rex, D.C. 1971.  Potassium-argon ages of slates and
    phyllites from south-west England.  Quart. Journ. Geol. Soc. Lond. 126,
    465.
Edmonds, E.A., McKeown, M.C. and Williams, M. 1969.  British Regional Geology
    South-West England.  3rd. Ed.  H.M. Stationery Office, London.
George, T.N. 1969.  British Dinantian Stratigraphy.  Compt. Rend. 6me. Congr.
    Strat. Carb. 1, 194.
Johnson, G.A.L. 1973.  Crustal Margins and Plate Tectonics during the
    Carboniferous.  Compt. Rend. 6me. Congr. Strat. Carb.
Matthews, S.C. 1974.  Exmoor Thrust? Variscan Front?  Proc. Ussher. Soc.
    3, 82.
Owen, T.R. 1974.  The Geology of the Western Approaches.  In:  The Ocean
    Basins and Margins.  Vol. 2.  The North Atlantic, (Ed: A.E.M. Nairn &
    F. G. Stehli).  Plenum  Press, New York.
Ramsbottom, W.H.C. 1969.  The Namurian of Britain.  Compt. Rend. 6me. Congr.
    Strat. Carb. 1, 219.
Ramsbottom, W.H.C. 1973.  Transgressions and Regressions in the Dinantian:
    A new synthesis of British Dinantian Stratigraphy.  Proc. Yorks. Geol. Soc.
Rayner, D.H. 1967.  The Stratigraphy of the British Isles.  Cambridge
    University Press, Cambridge.
Renouf, J.T. 1974.  The Proterozoic and Palaeozoic Development of the
    Armorican and Cornubian Provinces.  Proc. Ussher Soc. 3, 6.
Sadler, P.M. 1974.  An appraisal of the "Lizard-Dodman-Start Thrust" concept.
    Proc. Ussher Soc. 3, 71.
Walkden, G.M. 1972.  The mineralogy and origin of interbedded clay wayboards
    in the Lower Carboniferous of the Derbyshire Dome.  Geol. Journ. 8, 143.

CHAPTER 6

# "From New Red Sandstone Deserts to Chalk Seas"

In the previous chapter it was noted how the British area had begun to cross into the Northern Hemisphere after millions of years south of the Equator. By the early Permian, the Palaeo-Equator ran from Southern Newfoundland to N.W. Spain and Southern France with the $10^{\circ}N$ latitude through about southern Scotland (according to Van der Voo and French, 1974). The North Pole was south of the Kamchatka Peninsula and the South Pole was moving to Antarctica. More of the present-day Northern Hemisphere lands were moving into that hemisphere so that the "Arctic Ocean" was becoming more and more enclosed by land areas as the Carboniferous moved into the Permian. This, together with the movement of Britain away from the Equator (to almost $20^{\circ}N$ by the Permo-Triassic boundary), might well account for the change in climate from the steaming moist forests of the Westphalian to the hot desert climate of the Permo-Trias. Pressure patterns must have been influenced by the mass movements of large continents and by Permo-Triassic times the British area was firmly in the "Horse Latitudes" with mainly NE to E winds predominating, according to the sedimentological evidence (dune-bedding, grooving, sculpturing, etc.).

In the Triassic, Britain continued to move northwards (probably with not much longitudinal change) reaching $30-40^{\circ}N$ by late Triassic times and perhaps even to $40-45^{\circ}N$ by the early Jurassic (again according to Van der Voo and French). By late Triassic times the Palaeo-Equator had reached West Africa and was almost at the present day position at the beginning of the Jurassic. The North Pole during these earlier Mesozoic times was moving around in the East Siberian Sea getting to the tip of Alaska by the later portion of the era. The South Pole was hovering close to Antarctica and Tasmania. In the early Cretaceous there appears to have been a movement of Western Europe back south again with Britain reaching latitudes less than $40^{\circ}N$ in the Upper Cretaceous before finally continuing with its northward drift from early Tertiary times onwards.

The mention of drift raises of course the most important feature of the Mesozoic (and post-Mesozoic) continent relationships. This was the time of the great split of Pangaea, this was the continental drift envisaged by Alfred Wegener. Splits began to occur between Africa and the Americas as early probably as the Triassic. India began her rapid dash north. By Jurassic times, the split between West Africa and the eastern seaboard of North America was widening though very little widening was taking place further south in the incipient "South Atlantic". According to Smith, Briden and Drewry, the main South Atlantic spreading was delayed until late Cretaceous to Early Tertiary times when movement was rapid. By the end of the Mesozoic the tip of India had moved to about $25^{\circ}S$ and was to keep on pushing northwards.

The climatic changes over the British area in the Permian-Cretaceous time span fit the general movement northward from a dry hot belt to a wetter area in the Jurassic (though not necessarily a cooler one as summers may have been hot and monsoonal with northern and central Europe on the northern edge of a

101

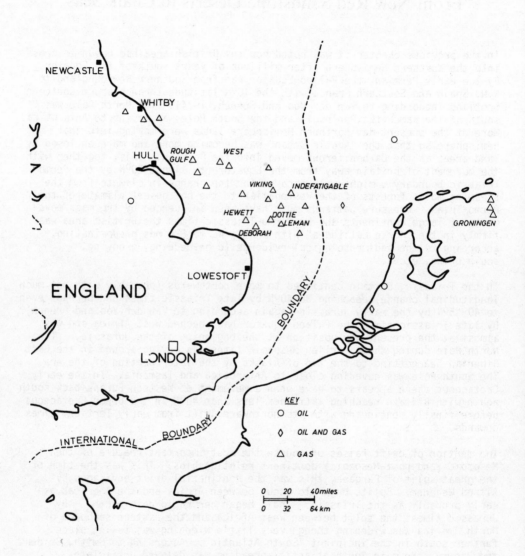

Fig. 56. The British Gas Fields in the Southern North Sea
West - West Sole Field

Tethys Ocean - a position like South China today).  Traditional beliefs think
of "desert" land areas surrounding the Chalk seas.  It may be that the
temporary southward retreat of Britain (and its environs) in the Cretaceous
brought it back into a drier zone.  On the other hand, the opening of the
(southern) North Atlantic would have encouraged a new moist source and it
could be that the lack of detritus into the British Chalk Sea is to be
mainly attributed to the low (eroded) level of the source areas rather than
to a desert position.

The Permian story in Britain can be compared, to some extent, with the
Devonian in that both periods followed major orogenic phases.  The Permian
story is largely one of erosion - the wearing down of the high relief produced by the Variscan (Hercynian) climax, just as the Devonian was the time
when the Caledonian mountains were eroded.  The Permian is therefore the
result of things that had happened.  The Mesozoic in Britain, however, is
really the story of "things to come", the story of the opening of the North
Atlantic.  The final real opening of the ocean happened between Greenland and
Norway, Greenland and Rockall, and between North America and the Porcupine
Bank, but there were many other futile (or semi-futile) attempts, abortive
openings along a number of deep seated fractures above regions of upwelling
mantle.  The main areas involved were the Rockall-Hatton Trough, the Rockall
Trough, the Minch-West Shetland trench, the Porcupine Bight, the Irish-
Celtic seas and the North Sea.  In the latter area there were a number of
"failed arms", the main ones being the Viking and Central grabens (see
fig. 57).  Whiteman (1975) refers to a 1200 km long system of mainly sub-
Upper Cretaceous troughs in the North Sea and sees them as a  superficial
aspect of deep-seated crustal structures in an overall context of trilete
trough systems.  Their development, as failed arms over crustal uplifts
during late Carboniferous to early Permian times, allowed thick accumulations
of Mesozoic and Tertiary sediments.  (Fig. 9 of "Geology of the North-West
European Continental Shelf, Vol. 1, 1975, gives the position of the trough
systems around the British Isles).

During Permian to end Cretaceous times then, important cracks were occurring
above potential spreading axes in the areas around Britain, and, in
attempting to spread, rifted graben systems were forming, some more successful at rifting (if not at spreading) than others.  It is to be expected that
all this would have more widespread effects, especially in terms of regional
uplifts, tilts, submergences and emergences as sea levels rose or fell and as
waters entered into and passed along the rifts.  Major advances of the sea
occurred at certain specific times in the late Permian and Mesozoic history
of the British (and surrounding) areas and it is significant that these
episodes are recorded also in the countries around the North Sea and elsewhere
in Western Europe (and even beyond).  The first of these advances was the
temporary one by the Zechstein waters in the later Permian, the flooding
beginning with the shallow spread in which the Marl Slate ("Kupferschiefer")
was thinly deposited.  These waters entered the North Sea from Poland and
from the north (through a breached barrier) but became cut-off with shrinkage and saline deposition.  The next advance was in Mid-Triassic times but
was more important on the Continent with the deposition of Muschelkalk. There
could have been some slight influence on the British depositional areas, with
the deposition of the Waterstones and with entry of water via the Celtic Sea
direction.  Then came the Rhaetic advance heralding the truly marine Lias
waters.  This was a major transgression and must point to important rifting

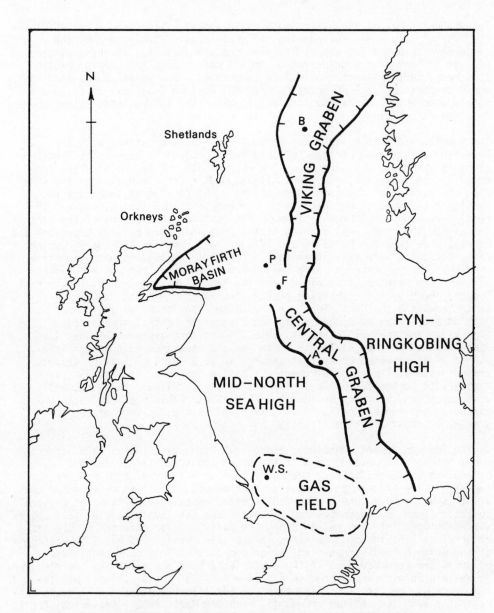

Fig. 57.  Structural elements in the North Sea (after Kent, 1975).
B. Brent; P.Piper; F. Forties; A. Argyll; W. S. West Sole.

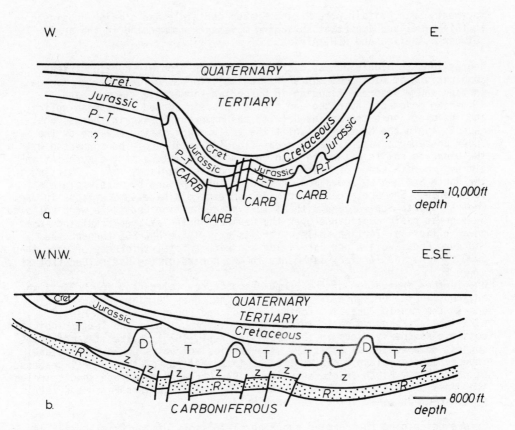

W.                                                                          E.

a.

W.N.W.                                                                  E.S.E.

b.

KEY:  P-T   Permo-Trias
        T    Triassic
        Z    Zechstein

        ▦    Rotliegende Sands
        D    Salt diapirs

Fig. 58.  Diagrammatic sections across the North Sea
(after Evans and Coleman, 1974) (a) across central North
Sea,  (b) across southern North Sea.

movements near Britain (the Viking Graben certainly had a major subsidence at this time) and important spreading movements commencing in the area between N.America and N.W. Africa.

Regressions in Middle Jurassic times were again probably due to various combinations of block movements (Kimmerian), to be followed by new marine inroads and widespread flooding in Oxfordian-Kimmeridge times. Late Kimmerian movements near the end of the Jurassic caused widespread uplift, and incipient unrests continued (with the numerous breaks in successions, especially in the North Sea) until the truly major marine advance by the Upper Cretaceous waters. This great flooding must surely be connected with the widening splits extending into more northern waters of the North Atlantic and up both sides of Southern Greenland (see Harland, 1969). Rifting in the North Sea had by now almost ceased, being replaced by more wholesale subsidence. End Cretaceous (Laramide) unrests did have some effects in the Central (North Sea) Graben with slump-type deposition but these were followed by a great basin subsidence over the Central North Sea area resulting in a great elliptical Tertiary pile, with its axis lying NNW-SSE through the Maureen, Cod, Ekofisk and Dan fields and with a total thickness of more than 3000 m of marine Tertiary sediments in the centre of the basin (Cod-Ekofisk).

One further indication of the Permo-Mesozoic crustal unrest in the British environs is the presence of volcanics in the Lower Permian (Denmark Sector) and in the Middle Jurassic (Forties Field) and Upper Jurassic (Frigg.). Alkali basalts and volcanic tuffs occur, distribution of the latter indicating possible southerly winds in the Jurassic (Kent, 1975). Laterites and boles suggest some subaerial activity. The Forties volcanics could be fracture induced as the volcanic centre lies at intersections of the central grabens with the Moray Firth fracture systems. On a broader scale one should include the hints of volcanic centres south of the English Fullers Earth and the presence of Jurassic intrusives in N.Brittany.

Before describing the Permian and Mesozoic history of the British area in greater detail, it is well to pause and consider the tremendous impact to British (and N.W. European) stratigraphy of the recent investigations in the North Sea. Fig. 56 and Fig. 69 show the positions of the main gas and oil fields in the North Sea. Broadly speaking, the gas fields are mainly in the south (fig. 56) extending in a WNW-ESE orientated oval (fig. 57) from E.Yorkshire and Lincolnshire out towards the Dutch coast. West Sole, Viking, Indefatigable and Leman are notable centres. In these fields the natural gas has migrated upwards from underlying Coal Measures into the Rotliegende Sands (Permian) which form the main reservoir. In Indefatigable, the Rotliegende is a massively-bedded, almost pure fluvial sand. In West Sole there is a three-fold division of the unit. The lower portion, 125 m thick, shows upward transition from aeolian to reworked to shallow water sands. The Rotliegende sands are capped by Zechstein dolomites and salt. Halokinesis of the Zechstein evaporites took place from Triassic times onwards forming numerous salt diapirs. There is often no Jurassic between the Trias and the Cretaceous, the absence of the system being attributable to Kimmerian uplifts and erosion over salt risings. Fig. 58(b) is a composite section across the southern gas field area to show the general relationships and broad structure. The Tertiary is less than 1000 m thick.

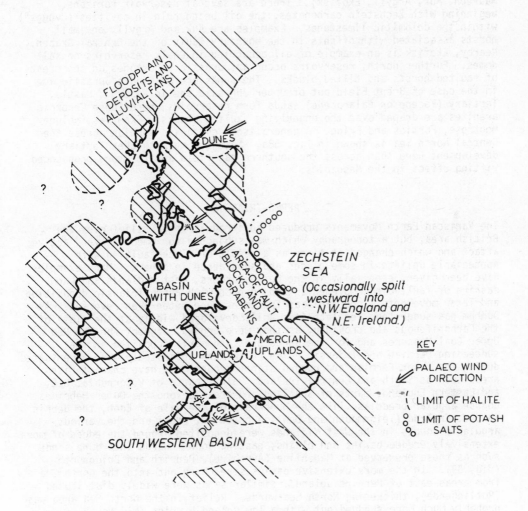

FLOODPLAIN
DEPOSITS AND
ALLUVIAL FANS

DUNES

?

?

ZECHSTEIN
SEA
(Occasionally spilt
westward into
N.W.England and
N.E. Ireland.)

BASIN
WITH DUNES

AREA OF FAULT
BLOCKS AND
GRABENS

MERCIAN
UPLANDS
UPLANDS

?

?

DUNES

SOUTH WESTERN BASIN

KEY

PALAEO WIND
DIRECTION

LIMIT OF HALITE

LIMIT OF POTASH
SALTS

Fig. 59.  Palaeogeography of Permian times.

The majority of the remaining North Sea fields yield mainly oil (fig. 69).
The majority lie near to the international mid-line (Brent, Frigg, Beryl,
Maureen, Auk, Argyll, Ekofisk).  There are several reservoir horizons,
beginning with Zechstein carbonates, the oil being held in cavities ("vuggs")
within the dolomitic limestones.  Examples are Auk and Argyll, on small
horsts associated with offsets in the boundary fault of the Central Graben.
Nearby, Ekofisk is an example of oil occurring in Chalk reservoirs on salt
domes.  Further north, reservoirs occur in Jurassic sandstones on the crests
of faulted horsts and tilted blocks.  The sands are of Middle Jurassic age
in the case of Brent Field but of Upper Jurassic age in Piper.  Lastly,
Tertiary (Eocene or Palaeocene) sands form reservoirs where these Cenozoic
arenites are draped over the underlying faulted blocks.  Examples include
Montrose, Forties and Frigg.  A generalised, composite section across the
Central North Sea is shown in fig. 58a.  Note the much thicker Tertiary
development here than across the southern gas field and the more pronounced
rifting effect in the Mesozoics.

## THE  PERMO-TRIASSIC

The Variscan Earth Movements produced an irregular high relief over the
British area, but a topography which was subjected to immediate very active
attack and which changed at times as a result of block-faulting with
appreciable uplifts in some places and subsidence in others.  Many examples
have been given, especially in the Welsh Borders, West Midlands and Lancashire of fault-block movements keeping pace with Permo-Triassic deposition
and later movements along Variscan fractures were often even reversed.  As
Dunham has suggested, red lateritic soils developed widely at the close of
the Carboniferous and provided, as they were eroded, pigment for the Red
Upper Coal Measures and for the sandstones and marls deposited in the
succeeding Permian and Triassic.  The relief was undoubtedly more accentuated
during the Lower Permian and rugged stoney ground must have characterised
areas such as South and Central Ireland, Wales, parts of W.Cornubia, S.E.
and Eastern England, parts of Scotland and areas beyond the Outer Hebrides.
Coarse angular breccias, like the brockrams of the Vale of Eden, the debris
in S.W. England (Teignmouth and Dawlish breccias, etc.) and the various
angular products in the West Midlands were deposited around the edges of more
extensively eroded basins and plains, passing soon inwards into dune sands
such as those preserved at Mauchline (Ayrshire), Penrith and Bridgnorth
(fig. 59).  In the more extensive basins stretching out into the North Sea
from areas east of Pennine uplands, similar sands were widely distributed
(Rotliegende), thickening North Sea-wards.  Relief in the North Sea area was
probably much more subdued but with a low upland barrier (Mid North Sea-
Ringkobing High) separating two basins (which may even have been below sea
level).  The basal yellow sands of Durham to Nottingham (fig. 60) belong to
this Rotliegende phase.  These North Sea basal sands rest on Carboniferous in
southern parts of the North Sea but probably on pre-Carboniferous (mainly
Devonian) basement farther north.  The Upper Permian began with a breach of
barriers letting in a boreal ocean in the North Sea, the mid-North Sea
barrier being (fault) breached at the same time.  Waters entered also from
the east, into the eastern (Polish) end of the Zechstein basin.  The basal
sands, originally of aeolian origin, were resorted and redistributed in
water.  A brief period of anoxic conditions like those of the modern Black
Sea has been suggested by Dunham (1974) to account for the thin black Marl
Slate ("Kupferschiefer").  Algal reefs developed around the western margins

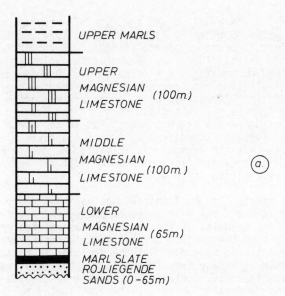

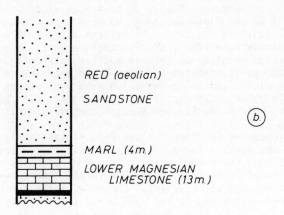

Fig. 60.  Contrast between Permian successions at
(a) Durham and (b) Nottingham.

of the Zechstein Sea while out to sea the rate of evaporation began to exceed
the addition of water supplied by rivers (Dunham, 1974).  Evaporation products began to form in order:  calcite (limestone), anhydrite or gypsum,
halite and bittern salts.  Three such evaporation cycles (the second and
third reaching the bittern stage) are found in S.E. Durham and N.E. Yorkshire
(fig. 60a).  The salt deposits were sealed in by the thick overlying red
marls.  The "Magnesian" limestones were formed by sea-bottom dolomitisation
of the limestone members of the cycles.  West of the Pennines, evaporite
cycle phases above anhydrite are not seen.  Out in the North Sea, then
dominated by the tideless Zechstein Sea, 4 cycles are known.  Olson and Ley-
den (1973) believe that the Permo-Triassic salts are due to concentrated
brines from the Mediterranean flowing into the Gulf (of Mexico) area and
along a volcanically active North Atlantic rift.

Shrinkage of the margins of the Zechstein Sea has been amply demonstrated by
Hollingworth and others.  The disappearance of higher Zechstein waters from
Nottinghamshire, for example, is seen in the return of red aeolian sandstones
(fig. 60b), for a time the subject of considerable controversy concerning
their Permian or Triassic age (not that the decision is clear nowadays).
Before its ultimate shrinkage away from Britain, the Zechstein water had
penetrated through into N.E. Ireland and Lancashire.  In the Irish Sea, the
equivalents of the Manchester Marls pass through shales to salts in a depo-
centre.

Difficulties of separating Permian from Triassic breccias, conglomerates,
sands and marls make the precise dating of separate events difficult in
certain deep basins in Britain. Bott has postulated that there is up to 2 km
of Permo-Trias in the Stranraer Basin, up to 4.9 km of Carboniferous and
Permo-Trias in the Solway Firth Basin and up to 6 km of Carboniferous-
Triassic in the East Irish Sea Basin.  Steel and Wilson (1975) claim that the
4000 m thick Stornaway Formation (largely conglomeratic) is of Permo-Triassic
age and was deposited on alluvial fans as mudflow, streamflood and braided
steam deposits, and on floodplains as channel and overbank sediments.  Phases
were tectonically controlled.  The formation is a sedimentary fill within the
deep western margin of an asymmetrical North Minch Basin in Permo-Triassic
times.  The Minch Fault (fig. 62) had a major controlling effect.

Undifferentiated "New Red Sandstone", i.e. Permo-Triassic, is now known to
occur widely off the south side of Cornubia beneath the waters of the western
Channel.  Inland, too, there may be thick "New Red" deposits (the Triassic
underlying the Isle of Purbeck may be 1500 m thick;  there may be thick
sequences beneath the Hampshire Basin).  The Permo-Trias thickness in the
Western Approaches Basin could exceed 2000 m and the red beds probably contain salt horizons.  Only a thin Triassic layer (250 m?) may occur in the
Bristol Channel, but this thickens eastwards into Somersetshire where the
Puriton Borehole showed 400 m of Triassic marls and salts overlying 320 m
of Triassic to Permian sandstones and conglomerates.

Permian volcanics on present day land in Britain are restricted to the Exeter
area (now that the Mauchline volcanics in Ayrshire are claimed to be of Upper
Carboniferous age).  The Devon volcanics extend, as small outcrops, as far
north as Loxbeare and well westwards along the Crediton Trough.  The well-
scattered remnants must point to once extensive flows of the lavas.  Small
areas in the Plymouth region may also be of Permian age.  The volcanic suite
comprises lamprophyres and olivine basalts, together with vent agglomerates.

The base of the Triassic in Britain is now drawn at the entry of the Bunter
Pebble Beds, so that the dune sandstones below this horizon are now
considered with the Permian.  The suggested broad framework at the time of
influx of the coarse rudites is shown in fig. 61.  The downfaulting Worcester
Graben was directing the northward transport of these river-borne gravels,
and correlates also with the first rifting attempts along the Central North
Sea Graben.

For details of the Triassic Rocks of the British Isles, the reader is
especially referred to the Symposium published by the Geological Society of
London (Vol 126, 1970).  Of particular interest are the excellent reconstructions of Triassic palaeogeography by Audley-Charles.  He subdivided the
Triassic into six lithostratigraphical divisions and presents palaeogeographical maps for each division (subdividing further the first into that
for the basal pebble beds and another for the overlying Upper Mottled Sandstone.  He showed that the grabens and most of the principal basins in which
the sediments accumulated were structurally controlled.  Triassic faulting
had an important influence on the thickness and nature of the sediment
deposited but the faulting waned towards the end of the Triassic and almost
ceased by the beginning of the Rhaetic.  Faulting also periodically caused
the floor of the northern end of the Worcester Graben to subside more than
the southern end making that rift the principal channel through which passed
much of the sediment deposited in central and northern England.  In this
way, according to Audley-Charles, the Midlands basin was supplied with eroded
debris from the Variscan uplands throughout most of the Triassic.

The Triassic climate, of the British area, was hot with wet and dry
alternating seasons.  Sedimentary structures show the presence of rivers and
lakes and water was abundant, at least periodically.  With the gradual lowering of relief, extensive bodies of standing water were established in which
the Waterstones (division 3) and the Keuper Salt (division 4) were deposited.

The relief was relatively rugged at the beginning of the Triassic.  Huge
braided rivers flowed between the uplands during the deposition of the Bunter
pebble beds.  Many of the pebbles are far-travelled and appear to have been
eroded from French Variscan highlands.  Slower flowing rivers probably
account for those water-deposited varieties of the overlying Upper Mottled
(St. Bees) Sandstone;  other variants are aeolian in origin, especially in
S.W. England.  Scottish equivalents are of fluviatile and lacustrine origin.
The Lower Keuper Sandstone (division 2) presents a number of facies.  The
nature of a pebbly facies in Nottinghamshire suggests more nearby sources
than the far-off south of the Bunter rudites and that there was a reworking
of those underlying gravels.  The large river from the south continued to
flow through the Worcester rift and carried its coarse load into the Cheshire
Graben and out into the nearby Irish Sea.  Over extensive areas however,
fluviatile sands were deposited.  Salts were forming off E.Yorkshire and over
the southern North Sea.

The succeeding Waterstones Formation (division 3) comprises shales, siltstones
and sandstones, totalling up to 60 m thick.  Audley-Charles believes the
Waterstones to be the British correlative of the marine transgression which
deposited the Muschelkalk on the Continent and in the southern North Sea.
Warrington has found marine microplankton in the Warwickshire Waterstones
while Rose and Kent found Lingulids in Nottinghamshire in sediments claimed
by Klein to be intertidal.  The Waterstones contain the first signs of salt

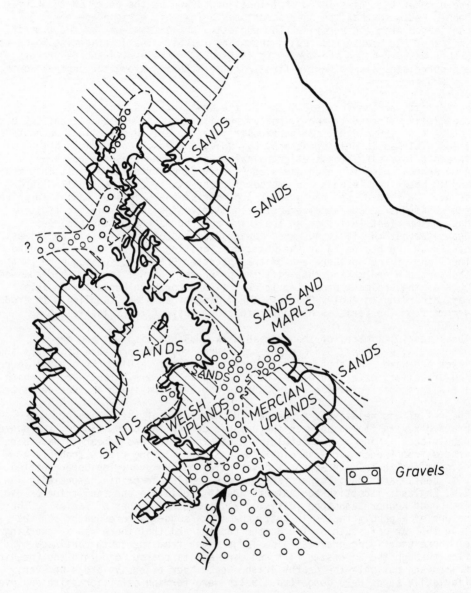

Fig. 61.  Palaeogeography of early Triassic times (in
part after Open University Series S23-Block 6).

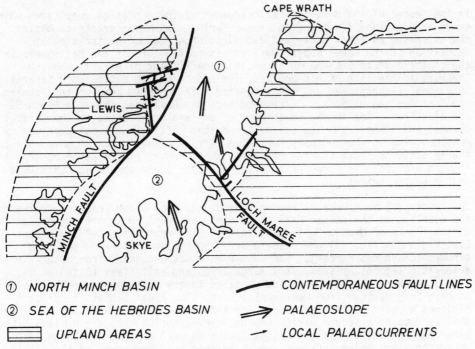

① NORTH MINCH BASIN            CONTEMPORANEOUS FAULT LINES
② SEA OF THE HEBRIDES BASIN    PALAEOSLOPE
UPLAND AREAS                   LOCAL PALAEO CURRENTS

Fig. 62.  Permo-Triassic basins in N.W. Scotland (after
Steel and Wilson, 1975).

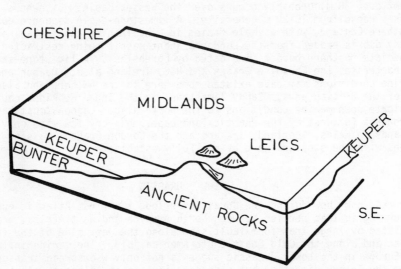

Fig. 63.  Overlap of Keuper on to the ancient rocks of
Charnwood Forest.

in the Keuper of the Midlands, a forerunner of the important evaporites which
occur in divisions 4 and 5 (the Keuper Marl).  Two major evaporite basins,
characterised by thick rock-salt deposits, occurred during the time of
deposition of the Lower Keuper Marl (division 4).  One of the basins was in
the southern North Sea and on to South Denmark.  The other extended from
Cheshire to the Isle of Man and N.E. Ireland.  The two basins were separated
by a zone of sandstones and silty mudstones in the area between the Pennine
and East Anglian uplands.  Brine movements may have been from the North Sea
and also from a sea to the south-west of Britain encroaching on the Bristol
Channel and South Irish Sea areas.  Important rifting and subsidence was by
now occurring in the northern North Sea (Viking Graben).  Earth movements in
the Triassic may have started halokenesis of Permian evaporites exposing
them to erosion.  This reworking of older salts could have also contributed
to the Keuper brines.

The Keuper was much more extensively deposited than older Triassic sediments
and overlaps of the Keuper Marl on to older basement rocks are common, as
on the flanks of Mendip "uplands", Vale of Glamorgan "islands" and across the
East Midlands (fig. 63) where the underlying buried topography has been
exhumed (Charnwood Forest).  The Keuper Marl was deposited from water - from
evaporitic lagoons, playas, very large lakes and salt flats (Fitch et al,
1966).  Much of the silt and sand supplied to the N.Irish Sea Graben appears
to have been derived from Western Scotland.  The Arden Sandstone of the
Midlands appears to be of estuarine or intertidal origin. Warrington believes
that it records a second marine or semi-marine influx into the Midlands, this
time from the south (via the Wessex Basin) as well as from the North Sea. The
Waterstones and Arden influxes are to be seen as forerunners of the really
widespread and important marine transgression which deposited the Rhaetic and
brought the Triassic period to a close.

In the southern North Sea, the Rhaetic (Winterton Formation) reaches 60 m in
thickness, greenish-grey sandstones and mudstones with a marine fauna.
Further north sandy deposition prevailed.  In Britain the greatest thickness
(just over 30 m) probably occurs over the Wessex Basin.  Elsewhere thicknesses range from 10-20 m generally.  A limestone-shale sequence dominates in
Southern England, with a shale facies in Lincolnshire and Yorkshire, with
sandy debris eroded from the London-Brabant massif being restricted to its
immediate northern border (indicated by boreholes).  A limestone-shale facies
characterises the Cheshire Graben and N.E. Ireland also.  Deeper and open
marine conditions may have existed somewhere to the NW here and also to the
SW of the British area.  Sandy limestones in the Inner Hebrides could
indicate open marine conditions to the north also.  Lagoons must have been
important features of the Rhaetic landscape, now very low with only gentle
uplands in Wales, Scotland, Ireland and the London-Brabant massif.  Connections with the open sea were by now fully established.


THE  JURASSIC

Fig. 64 shows how during the Jurassic Period, the first Atlantic opening was
occurring in that portion between North America and West Africa, probably
assisted by large transform fault slip along the west side of the Tethys
Ocean and along the Gold Coast-South America join.  The marine influx into
NW Europe in the Lower Jurassic suggests not only widespread transgression
from the Tethys direction but also some inroads by Atlantic waters along
tentative splits up the western side of Europe.  Fig. 65 shows a possible

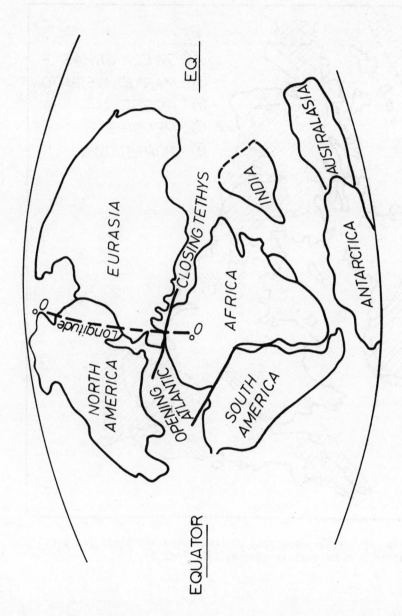

Fig. 64. Break-up of Pangaea by the end of the Jurassic (from The Breakup of Pangaea by R. S. Dietz and J. C. Holden. Copyright © 1970 by Scientific American, Inc. All rights reserved)

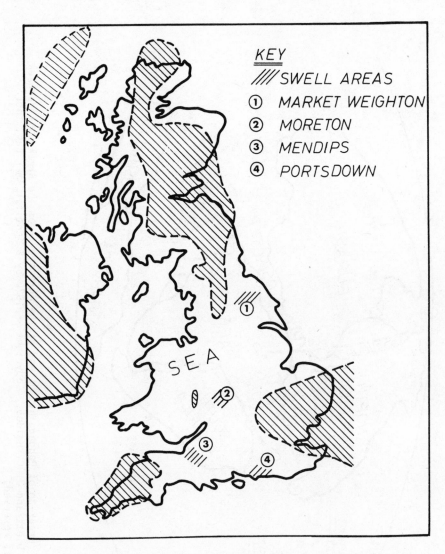

Fig. 65.  Palaeogeography of Lower Jurassic times (in part after Open University course, S23-Block 6).

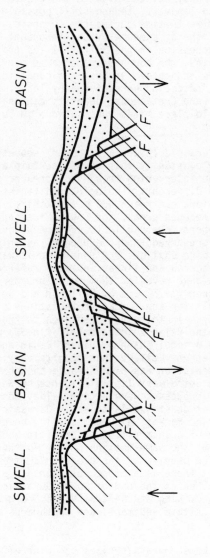

Fig. 66. Effect of basement fracturing on later sedimentation producing basins and swells (after Whittaker, 1975).

palaeogeography for the British area in Lias times.  The main non-
depositional uplands were probably low in height and included the East
Anglian-Brabant block, parts of Cornubia, parts of Ireland (though possibly
much less than is traditionally marked out) and a long whaleback extending
north from the Pennines into Scotland and beyond.  The Scottish uplands
were probably higher and of more varied relief than the remainder of the
British hills.  Some islands probably existed also to the west of the
Hebrides (Hudson, 1964).  A Welsh landmass, though still possible as a
rejuvenated shoaling area from time to time during the Jurassic, is no
longer really tenable in view of the disclosure by the important Mochras
Borehole of the thickest Lias succession in the British Isles (fig 67).  Here,
on the west coast of Merionethshire, the borehole penetrated over 1300 m of
Lias mudstones with marine faunas indicative of open-sea conditions.  One
would surmise that there was no indication here of nearby land.  On the other
hand, the way in which Jurassic (and Cretaceous) deposits can change very
rapidly in lithology from place to place makes one cautious in suggesting a
complete Welsh drowning.

One outstanding feature of the Jurassic (especially Lower) sequences in
England is the way in which they constantly thin across certain areas (shown
in fig 65), and thicken up in the intervening troughs.  Moreover the
attenuating areas often separate different lithologies and facies.  In the
thick trough areas, Lias successions can reach 460 m (Yorkshire) and 550 m
(Lower Severn Basin).  Over the reduced areas, the Lower Lias can be only
30 m thick at Market Weighton (where, due partly to Cretaceous overstep, only
that portion of the Jurassic is preserved) and the Upper Lias only 2 m thick
at one locality over the Oxfordshire shallows.  At one time it was thought
that these thinner regions were narrow and linear and were referred to as
"axes of uplift".  Later studies have revealed them to be broader "swell"
areas.  Recent studies by Whittaker (1975) and by Sellwood and Jenkyns (1975)
have suggested underlying controls for the thickness variations.  Whittaker
attributes the changes in thickness in the cover Mesozoics to fracturing
positive and negative blocks in the Palaeozoic basement (fig 66) with large
graben structures underlying the Bristol Channel, Severn Basin and the Weald
Jurassic trough.  Sellwood and Jenkyns believe that despite varying great
and negligible subsidence, sedimentation was always rapid enough to maintain
a roughly level sea floor, there being very little evidence, by way of slumps
or turbidites, that redeposition processes were active.  They too relate the
Mendip, London Platform and Dorset coast swells to early Jurassic positive
fault motions in the basement.  The Market Weighton "swell", however, they
ascribe to relative buoyant rise of a salt pillow or granitic body whose
movement was probably triggered by the same motions.  They further relate
these extensional tectonics to the opening of the oceanic "central" Atlantic
and Alpine-Mediterranean Tethys.  There is a close relationship also with
Kimmerian movements in the North Sea at the end of Lower Jurassic times,
causing widespread rifting in the Viking and Central North Sea grabens and
resulting in rapid infilling by deltaic sediments and southwards retreat of
the sea.

These movements at the close of Lower Jurassic times had wide effects,
especially on the sedimentation.  The largely open-water argillaceous (but at
first often carbonate-rich) conditions of the Lias were substantially changed
by the beginning of Middle Jurassic times.  Over southern and central areas
of Britain, muds gave way to the very shallow-water (almost emergent)
carbonate sedimentation of the Inferior and Great Oolite series.  Further

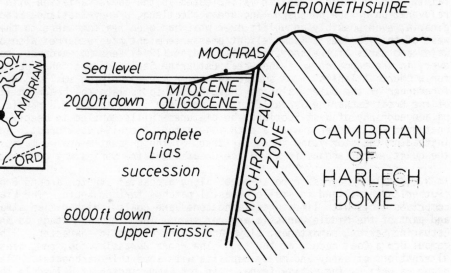

Fig. 67.  The Mochras (Llanbedr) Borehole.

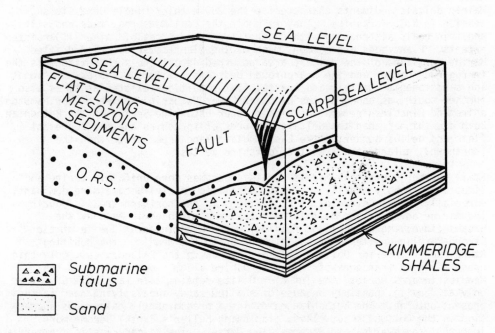

Fig. 68.  Reconstruction of Kimmeridge times in the
Helmsdale region (after Hallam, 1965).

north in both Britain and the adjacent North Sea, deltaic lagoonal and
estuarine sedimentation (almost reminiscent of the Coal Measures) set in.  A
fairly general northern uplift is indicated in the above variation with some
rejuvenation of the northern land areas (Scotland, Orkney-Shetland, Scandinavia, even possibly an uplift over what had been sea somewhere to the
west of the Hebrides).  Some slight emergence might have occurred also over
Wales and parts of Ireland (fig 70).  Hudson (1964) sees no evidence, however, in the sandstones of the Great Estuarine Series of the Inner Hebrides
for a "North Atlantis".  A source to the west need only be small.  "A
forerunner of the Outer Isles may have helped to isolate the brackish lagoons
of the Great Estuarine Series from the open sea".  As to the M.Jurassic
palaeogeography of West Scotland, he pictures "hills mantled in deeply-
weathered soil, covered with a rich vegetation of coniferous forest;  and
instead of the stormy and irregular deeps of the present Minch we must see
the quiet, warm lagoons in which the Great Estuarine Series was deposited".

In N.E.Scotland, Jurassic rocks occur along a coastal stretch around Brora,
(Sutherlandshire) and in smaller coastal patches farther south.  The Lias
comprises clays with limestones, sandstones and coals.  Higher Lias strata
and part of the Middle Jurassic are not seen.  The Bathonian Stage is an
Estuarine Series, sandstones, clays and shales of cyclic character.  The
famous Brora Coal occurs at the top.  The Upper Jurassic, too, comprises
alternations of sandy and muddy deposits with some thin carbonates.  Plants
occur as well as the marine faunas.  In the Kimmeridgian at Helmsdale there
is clear and spectacular evidence of submarine faulting with a jumbled mass
of Old Red Sandstone boulders set in distorted Upper Jurassic clays and
shales (fig 68).

Mainly deltaic sediments characterise the 250 m thick Middle Jurassic succession in N.E. Yorkshire.  They resemble the Coal Measures, muds and silts
and thin coals alternating with channel-type sandstones of rather impersistent
extents.  Plant beds are common, much of the plant material being drifted.
Marine invasions of the deltaic area occurred depositing such horizons as the
ferruginous Dogger and the Scarborough Beds (fossiliferous shales, limestones
and sandstones).  Deltaic and coastal flat conditions existed at times also
further south, as shown by the Bajocian Lower Estuarine Series of Northampton-
shire and Lincolnshire and the higher Upper Estuarine Series of the Bathonian.
Both are thinner than the deltaic measures of Yorkshire and may represent
flats and deltas bordering the East Anglian land area rather than the
(northern?) uplands that fed the Yorkshire deltas.

Still further south, the Middle Jurassic becomes more uniformly marine and
limestones (rubbly, shelly, oolitic, pisolitic, marly) become more dominant,
especially in the Cotswolds.  Oscillations in the northern areas producing
the marine advances and deltaic retreats  are mirrored in Avon by the
breaks (involving even slight folding) within the Inferior Oolite of the
Stroud-Birdlip-Cheltenham area.  Traced on to the Mendips, the Doulting
Stone (Upper Inferior Oolite) transgresses on to the Palaeozoics.  Only this
upper division transgresses into Oxfordshire also.  Southwards from the
Mendips towards Dorset, the Inferior Oolite remains thin (a couple of metres
only at Yeovil), possibly because of shallow sandy shoals lying near to the
present outcrop.  The Bathonian maintains a predominantly calcareous succession in the Cotswolds but clays, (including Fuller's Earth) become more
important southwards into Somerset and Dorset.  The volcanoclastic, commercial
clays may relate to a volcanic centre in the Western Approaches (Kent, 1975).

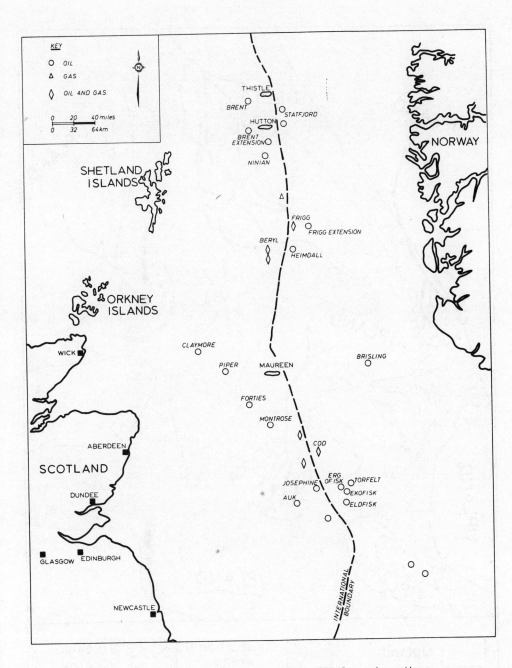

Fig. 69. Oil and gas fields in the middle and northern
North Sea. Argyll is to the southeast of Auk.

GEBI - E

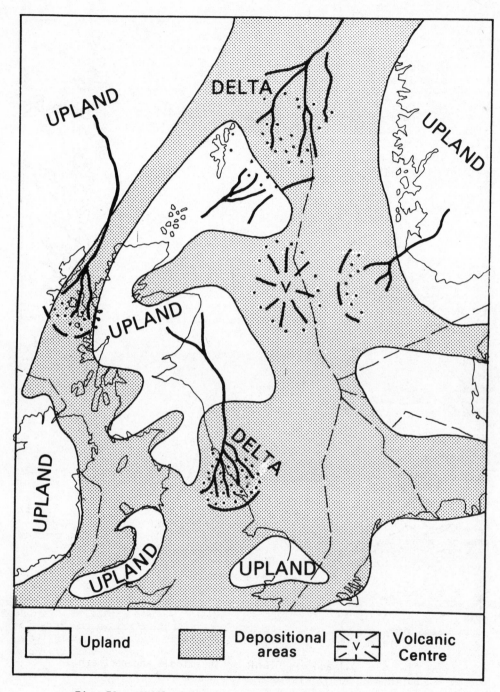

Fig. 70.  Middle Jurassic palaeogeography of Britain and the North Sea (after Kent, 1975).

The Upper Jurassic of England comprises a variable upward succession of clays (Callovian, Oxfordian and Kimmeridgian), limestones (Corallian, Portlandian) and sands (Kellaways Rock, Portland Sand). Marked oscillations of sea level are indicated, perhaps relating to various openings or closings of various marine inroads along possible spreading lines within and around the British area. The Corallian represents a marked shallowing between two important marine advances (the Oxford and Kimmeridge Clay). The higher of these two great clay formations reaches over 500 m in thickness in Wessex and the Weald. Marked marine retreats began to occur however at the close of Kimmeridgian times with emergence probably affecting large portions of the British area. The sea retreated to Southern and S.E. England and to a separated area embracing Norfolk, Lincolnshire and Yorkshire. The Wessex-Weald basin became largely a brackish to freshwater area by Purbeck times. It may have had westward connections via the Bristol Channel to a Celtic Sea depression in Portlandian and Purbeckian times. The Norfolk to Yorkshire area probably remained more permanently marine through the Jurassic-Cretaceous boundary.

Reconstruction of Jurassic times in the northern and central North Sea has been greatly facilitated by the search for hydrocarbons in recent years. Fig 69 shows the distribution of these main oilfields, British and Norwegian. The important discoveries, from a palaeogeographical point of view, of this search are of the important oil-bearing sands which occur at several horizons in the Jurassic and of the Middle Jurassic volcanics in and near the Forties oilfield (fig 70). The lowest sands occur in the Rhaetic-L.Lias of Brent Field (fig 71) and are 177 m thick (Kent, 1975). In the same area, 60 m of marine Lias shale are succeeded by the Middle Jurassic Brent Sands, 240 m thick and of estuarine-deltaic origin. The westward-tilted oil-bearing sands are shale-trapped by Upper Jurassic and Upper Cretaceous unconformities (fig 71). Kent believes the most likely source for the Brent Sands (containing coals) to be the Scottish Highlands and the far north rather than the Mid-North Sea High. The volcanic centre was in the Forties-Piper area. The lava flows (partly subaerial) are mostly undersaturated olivine basalts. Faunas interbedded with volcanic tuffs indicate a Bathonian age. The locations of the vents were probably controlled by the E-W faults of the "Witches Ground Graben", between Forties and Piper and the main Viking Graben to the east (see Kent, 1975, fig 9). The (Oxfordian) Piper Sands (oil-bearing) transgress the volcanics. The sands, up to 100 m thick, represent a high energy beach bar complex (Kent, 1975). Kimmeridgian marine conditions were again widespread and the shale represents a fairly sudden deepening, perhaps indicating some accentuated stretching of the main North Sea grabens. Some volcanic tuffs interbedded with the Kimmeridge Shale of the Brent area could relate to a S.E. Norwegian centre.

During the Jurassic, the northern and central grabens of the North Sea were probably stretching fairly continuously, but with some tilting of block members (especially along the graben margins) at times resulting in contemporaneous erosion of uptilted edges. One such important phase occurred after the end of Middle Jurassic times with a further slight movement before the Kimmeridgian. Important fault-movement and erosion preceded Lower Cretaceous deposition (see fig 71) and this phase may also have removed what was possibly a widespread Portlandian cover (present at Brent). In the Southern North Sea, Triassic and Jurassic deformation (and resulting erosion of different units) was largely due to salt tectonics.

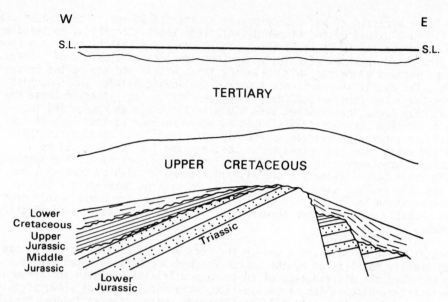

Fig. 71.  Brent Field (Kent, 1975, fig. 11).

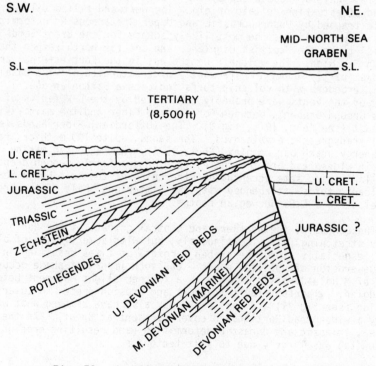

Fig. 72.  Argyll Field (Kent, 1975, fig. 16).

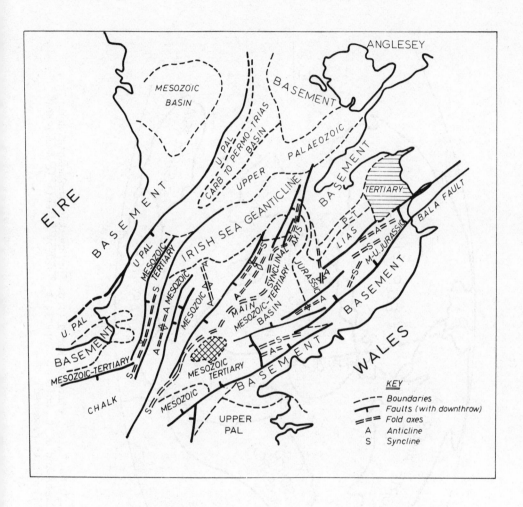

Fig. 73.  Sea-floor geology of the South Irish Sea (after
I.G.S. publication No. 73-11).  A north-south fault zone
should be indicated along the Merionethshire coastline.

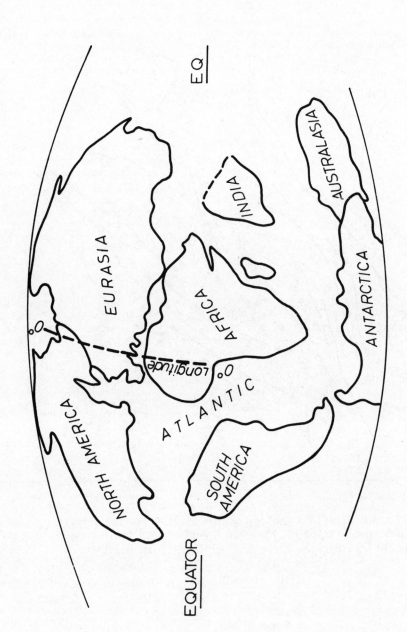

Fog. 74. The late Cretaceous World (from The Breakup of Panagaea by
R. S. Dietz and J. C. Holden. Copyright © 1970 by Scientific American,
Inc. All rights reserved)

## THE CRETACEOUS

It was during the Cretaceous that the main separation of South America from Africa took place (fig. 74) and with important splitting and spreading moving up between Western Europe and North America.  Roberts (1975) believes that Rockall Trough began to really open soon after the start of Cretaceous times (about 120 m.y. ago) but ceased spreading by the end of the period.  Pre-Oligocene faulting has preserved pockets of Cretaceous on the floor of the Rockall basin.  Bailey and Haynes (1974) too believe that Rockall Trough may represent the oldest part of the northern Atlantic (north of 51°N).  The Porcupine Seabight (west of Ireland) may include some late Jurassic to Lower Cretaceous sediments at its northern end (Dr. Bailey, personal communication).

Continental conditions affected a large part of the British area in lowest Cretaceous times, a result of the widespread uplift and regression towards the close of the Jurassic.  Uplifts were probably greatest in the northwest and west directing drainage southeastwards into a delta complex covering S.E. and Southern England (fig. 75).  Deltaic deposition also spread into the restricted areas of the Celtic Sea, the kind of Lower Cretaceous deposits in which the Marathon gas field (South of Ireland) occurs at depths of about 1000 m.  Allen (1967) has given a detailed reconstruction of the geology of the "London Uplands" during a part of Wealden times.  The edge of the Wealden deltaic basin was virtually the base of the gentle Middle and Upper Jurassic foothills, these sediments lying unconformably on a rising basement of Palaeozoics, getting older northwards and northeastwards (fig. 76).  Important rivers fed in from the London area and across the Dover-Margate portion of N.E. Kent.  Allen has demonstrated the cyclothemic nature of the Wealden, each cyclothem representing the southward pushing of a delta into a large freshwater lake followed by submergence under lake muds.  At times of maximum flooding the rivers became estuaries.  Iguanodon footprints demonstrate the shallowness of the water at times.  The western edge of the Wealden basin is suggested by the presence of pebbles of Cornubian-type cherts and tourmaline-rich rocks in Dorset.  The Dartmoor Granite appears to have been unroofed by this time.

In the marine waters of the North Sea, thin Lower Cretaceous muds were deposited, these waters extending on to the Norfolk-E.Yorkshire areas.  In Norfolk the deposits are coastal sands and muds, but in Yorkshire the Speeton beds are largely grey clays.  Deep water muds characterise the Danish portion of the North Sea, but these deposits fringe northwards and the Ringkobing-Fyn High was not covered until Aptian-Albian times.  Albian deposition in the southern North Sea, was like the Norfolk-Yorkshire fringe, of a Red Chalk facies.  In the northern North Sea, Neocomian deposition may be restricted and the transgressive basal Cretaceous (following on the late Kimmerian fault-movements) is mainly of Aptian-Albian age.  These deposits may rest on rocks as old as Triassic or even Devonian.  In the Piper Field, Lower Cretaceous carbonates pass up into thicker sands.  Thicker deposition still tended to occur within the main grabens and some contemporaneous tilting of blocks still continued in places (as at Argyll field, fig. 72).  Some vulcanicity has been located in the northernmost North Sea.

Fig. 75.  Early Cretaceous times in the British area.

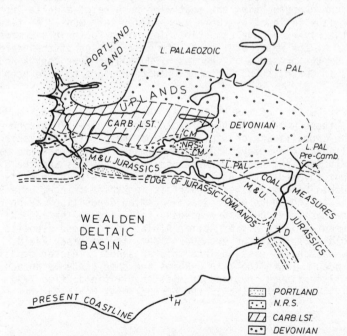

Fig. 76.  Palaeogeological reconstruction of the "London
Uplands" during part of Wealden times (after Allen, P.1967).

Possibly as a result of important sea-floor spreading into the Labrador Sea
(and along the Rockall Trough), a widespread submergence affected much of
Western Europe by about the beginning of the Upper Cretaceous, especially by
the beginning of the Albian.  Central and Northern England became submerged.
From the Rockall direction the sea then probably spread east into N.W.
Britain, including N.E. Ireland.  The Upper Cretaceous transgression (mostly
beginning with Albian deposits but in places being delayed until the
Cenomanian) is an important one in British stratigraphy, especially now that
the geology of the British seas is better known.  Moreover the transgression
was preceded by important lower to mid-Cretaceous movements involving in
places fairly tight folding and appreciable faulting.  Fig 79 shows the
warped and fractured floor (of the English Channel) on which the Upper
Cretaceous was deposited.  Note the rising Jurassic-Triassic-Permian succession in a westward direction.  This is dominant along the English coastal
sections from Dorset to Devon, where, on the Haldon Hills, Upper Cretaceous
sands rest on the Permian (fig 77a).  Fig 77b suggests how important downfolding might well have occurred also over the Bristol Channel area before
the deposition of the Upper Cretaceous.  The Bristol Channel downfold, now
preserving about 2000 m of Mesozoics need not be a mid-Tertiary fold structure
(faulting is a different matter).  It is tempting to suggest that Wales suffered an upwarping by mid-Cretaceous times.  Arkell has described the
particularly severe affects of pre-Upper Cretaceous deformation in the
Abbotsbury-Chaldon areas (fig 78), north of Weymouth.  The important Abbots-
bury Fault is largely a fracture of that age.  It is again tempting to
involve at least some westward downfaulting of Jurassics by the Mochras Fault
on the Coast of Merionethshire, in North Wales, at this time.

How much of the British area was really submerged beneath the Chalk Sea, at
its maximum transgression, will probably never be known.  The study of the
British seas has revealed how extensive the Chalk really was.  Moreover finds
such as that by Walsh in S.W. Ireland, show that even traditional Chalk
"islands" may not have been islands at all.  The Chalk is a remarkable
formation, being found in the Celtic Sea, English Channel, Southern North
Sea, Northern France, Denmark, N.Germany and southern Sweden, as well as in
our English and north Irish outcrops.  The Chalk, much of its finer material
derived from coccolith algae, was probably deposited in water of 200 to 300 m
depth.  Chalk variations must reflect subtle changes in water depth (usually
shallowings) and current activity.  Such pure deposits would seem to indicate
either very low-lying, peneplained land areas or very extensive waters with
only far-distant land sources.  Both causes probably apply.  There is after
all in, for example, the Cretaceous succession of the Weald, a gradual
fining-upwards sequence from the coarse Hastings sands to the fine-grained
Chalk indicating the long-term lowering of source areas.  The thickest Chalk
occurs in the Wessex Basin (over 600 m) and in Yorkshire (500 m), the
deposit being much harder in the latter area.  The highest preserved Chalk
(on land) occurs in N.Norfolk and in Antrim (the Pellet Chalk).  In N.E.
Ireland the lower portions of the Chalk are mostly missing, only the Senonian
Chalk having a wide distribution.  Where Turonian and Cenomanian deposits
occur (on the east coast of Antrim) they are mostly glauconitic sands with
intraformational breaks in the sequence.  Senonian seas probably had a wide
distribution also in the Scottish area.  Cenomanian beds are sandy in West
Scotland and are probably reworked.  Turonian is usually missing but the
preserved Senonian is chalky.  Only blocks are to be found on the eastern
coastal area and they probably come from the nearby North Sea floor.

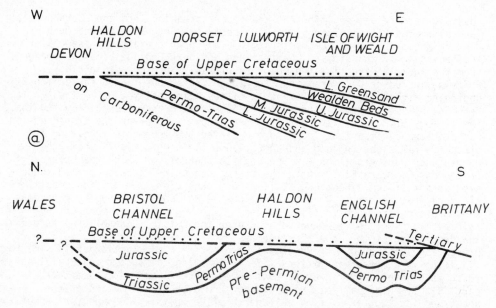

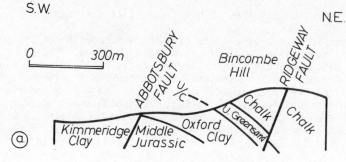

Fig. 77.  Upper Cretaceous overstep in South and South- West England.

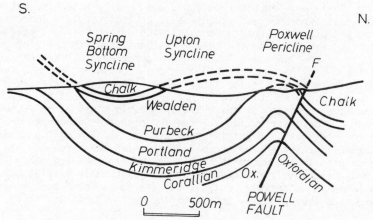

Fig. 78.  Mid-Cretaceous structures in Dorset (after Arkell, 1946).

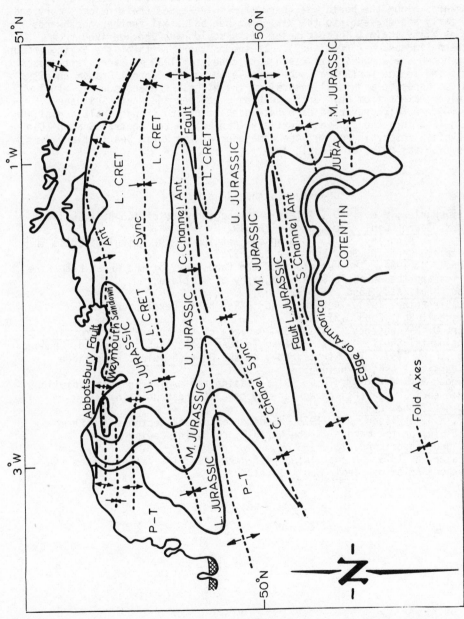

Fig. 79. Channel outcrops of pre-Upper Cretaceous rocks either on sea-floor or inferred beneath the sub-Upper Cretaceous unconformity (after Dingwall, 1971).

In the North Sea itself, the rifting had by now virtually ceased but subsidence was still greater over the sites of the grabens.  The Chalk reaches
900 m over the central graben.  Over the Fyn-Ringkobing High it is reduced
to about 250 m, and over salt structures in the south it is even less.
Northwards along the North Sea, hard (Yorkshire-type) Chalk becomes more and
more marly and shaley into the Viking Graben but still further northwards
becomes silty again.  These coarser silts could have been derived from
northern landmasses (Greenland?).  In the northern parts of the Viking trough
these Upper Cretaceous silts reach a thickness of 1000 m.  Very last phases
of rifting in the North Sea, more especially in the Central Graben resulted
in slump deposits and local removal of the highest Cretacous beds.  Inflow
of colder waters from the North (between Greenland and North Sea, see
fig. 80) probably prevented carbonate deposition and though Chalk deposition
was to last into the earliest Tertiary (Danian) in the Central Graben (Dan
and Ekofisk reservoir rocks), the remainder of the North Sea Tertiary
deposition was to be of clastic type.

## Suggested Further Reading

Audley-Charles, M.G. 1970.  Triassic Palaeogeography of the British Isles.
    Quart. Journ. Geol. Soc. Lond.  126, 49.
Dunham, Sir Kingsley, 1974.  Geological setting of the useful minerals in
    Britain.  Proc. R. Soc. Lond.  A.339, 273.
Hudson, J.D. 1964.  The Petrology of the Sandstones of the Great Estuarine
    Series and the Jurassic Palaeogeography of Scotland.
    Proc. Geol. Ass. Lond. 75, 499.
Kent, Sir Peter. 1975.  Review of North Sea Basin development.
    Jl. geol. Soc. Lond. 131, 435.
Naylor, D. & Mounteney, S.N. 1975.  Geology of the North-West European
    Continental Shelf.  Vol 1.  Graham Trotman Dudley Publishers Ltd. London.
Rayner, D.H. 1967.  The Stratigraphy of the British Isles.  Cambridge
    University Press, Cambridge.
Sellwood, B.W. & Jenkyns, H.C. 1975.  Basins and swells and the evolution
    of an epeiric sea (Pliensbachian-Bajocian of Great Britain).
    Jl. geol. Soc. Lond. 131, 373.
Steel, R.J. & Wilson, A.C. 1975.  Sedimentation and tectonism (? Permo-
    Triassic) on the margin of the North Minch Basin, Lewis.
    Jl. geol. Soc. Lond. 131, 183.
Whittaker, A. 1975.  A postulated post-Hercynian rift valley system in
    southern Britain.  Geol. Mag. 112, 137.

## CHAPTER 7

# The Final Moulding

The Cenozoic Era has so far lasted 65 million years. This is no more than
the length of certain periods, such as the Cretaceous or the Carboniferous,
and to further subdivide it into some six "periods" is bound to make each
subdivision very short when compared with the periods in the remainder of the
Phanerozoic. It seems sensible therefore to merely subdivide the Tertiary
into Palaeogene (about 40 m.y. in length) and the Neogene (about 25 m.y.
long). In fact it seems sensible to extend the Neogene to the present-day so
that it further subdivides into three portions - Miocene, Pliocene and
Quaternary (compare fig. 1).

By early Palaeogene times, Britain had moved northwards to occupy a position
between 40 and 47°N. The Equator was over Northern Nigeria and Gibraltar was
at about the Tropic of Cancer. The North Pole, after having been over the
N.W. tip of America had moved into the E.Siberian Sea once more. A wide
opening of the South Atlantic was continuing and the Antarctic continent was
moving southwards to be almost centred over the South Pole. India continued
its rapid movement north, moving through some 30 degrees of latitude in about
60 million years of time.

The main sea-floor spreading in the North Atlantic now spread to the area
between Greenland and Norway (and between Greenland and the British area
(fig. 81), but the spreading movements were probably complex. Vogt and
Avery (1974) see the complexity as being due to a variable discharge from a
mantle convection plume now under Iceland. Before Eocene times, the North
Atlantic really continued into the Labrador spreading area, but at about
60 m.y. ago a discharge peak split Greenland from Europe. The result was
the large basalt floods on the continental edges and south of Rockall Bank.
Brooks (1973) considers that Britain in Palaeogene times could have overlain
a mantle zone like that under Iceland today, a mantle zone of anomalously
low seismic velocity and density. Isostatic uplift (because of this
anomalous density) resulted in pre-basalt erosion and in stretching allowing
the upward invasion by magma. George (1965) has drawn attention to the movements (including appreciable faulting and warping) and widespread erosion
which preceded the basalt extrusion. Fig. 82 shows how the Scottish Tertiary
basalts rest on a floor of rocks varying in age from Moinian and Torridonian
to Jurassic and Cretaceous. The Camasunary Fault suffered a particularly
strong movement in post-Cretaceous to pre-basalt times. Similar impressive
movements took place in NorthernIreland. The intense removal of Cretaceous
is particularly important and significant when one attempts to unravel the
erosion history of other British areas, such as Wales or the Lake District.
It is really asking too much to suppose that Chalk covers could last well
into Upper Neogene times over areas like Wales when such covers had been
appreciably deformed and virtually removed from West Scotland and parts of
Northern Ireland before the beginning of Eocene times, some 40 to 50 million
years earlier.

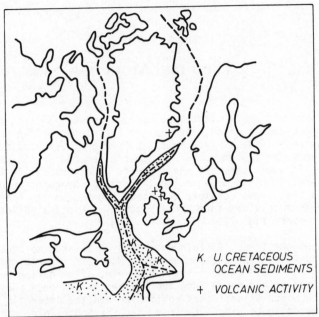

Fig. 80.  Minimum opening of North Atlantic in late
Cretaceous times (after Harland, 1969).

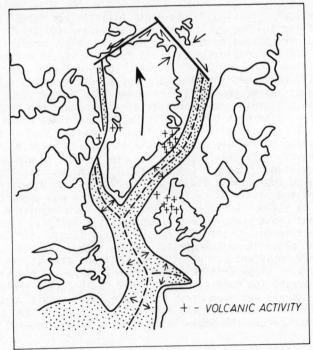

Fig. 81.  Early Tertiary opening of the North Atlantic
(after Harland, 1969).

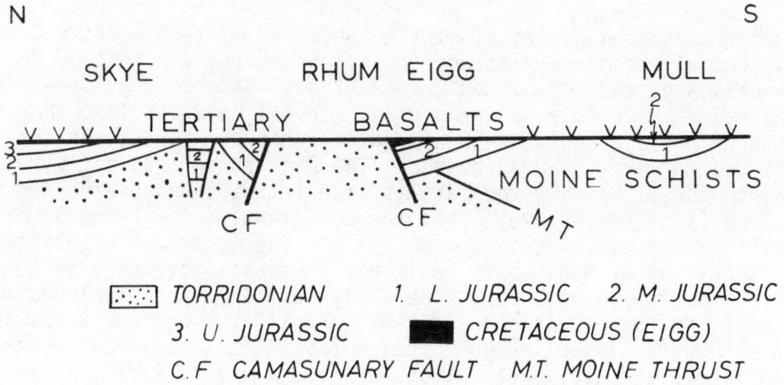

N                                                      S

SKYE            RHUM EIGG            MULL

TORRIDONIAN    1. L. JURASSIC    2. M. JURASSIC
3. U. JURASSIC    CRETACEOUS (EIGG)
C.F. CAMASUNARY FAULT    M.T. MOINE THRUST

Fig. 82.  Surface on which the Tertiary basalts were
extruded in W. Scotland (George, 1965, fig. 1.16).

1,2,3, = CENTRES (In order of origin)
MARGIN OF SUBSIDENCE
TERTIARY

Fig. 83.  The calderas and plutonic complexes of Mull and
Ardnamurchan (after Geological Survey guide to the model
of Ardnamurchan, 1934).  Lined shading represents Tertiary
basalts. G.G.F.  Great Glen Fault

The basalt floods in N.W. Britain lasted about 10 million years and were followed (but with some overlap) by the intrusive phase. Most of the igneous activity ceased by the end of the Eocene (though some 25 m.y. dykes have now been recorded from the Dingle Peninsula). Brooks believes that the "hot spot" under W.Britain had a NW-SE elongation in order to explain the linear extent of the former igneous province. The British igneous activity can be divided into three phases - the basalts, the intrusive complexes and the dyke swarms. The first, extrusive, phase was an extensive upwelling of basalts from Hawaiian-type volcanoes. The early Scottish lavas are associated with variable sandy, muddy and pebbly sediments. Forests and lakes must have been engulfed by the advancing lavas. The early Tertiary flora included magnolia, ginkgo, hazel and oak. The classic find in Mull of a charred tree, in position of growth, is well-known. Red boles indicate tropical weathering of the subaerial lava fields. The lavas were largely very fluid, individual flows being usually up to 15 m thick. Some 1500 square miles of lavas are preserved in Antrim and some 2000 metres thickness in Mull. The Irish flows comprise olivine basalts with a middle succession of tholeiites. Some trachytic flows occur in the Inner Hebrides.

The second igneous phase took the form of complex igneous intrusive centres. These occur in Skye, Mull, Arran, Ardnamurchan, Rhum, the Mountains of Mourne, Slieve Gullion and Carlingford. The 52 m.y. dating of the Lundy granite extends the area covered by these complexes. Brooks believes that the Lundy granite is associated with an underlying basic pluton and therefore compares in importance with the Scottish and Irish centres. Bott and Tuson (1973) suggest that the Skye and Mull complexes are each underlain by up to 3500 cubic kilometres of basic and ultrabasic rocks in the form of truncated cones extending to a depth of about 14 km if the composition is predominantly gabbroic. Substantial volumes of basics or ultrabasics also underlie the Ardnamurchan, Rhum, Mourne and Blackstones complexes. Bott and Tuson suggest that the principal rise of magma from the lower crust has taken place by piecemeal stoping and/or by cauldron subsidence.

The complexes comprise large numbers of concentric intrusions such as ring dykes and cone sheets. The ring dykes are of plutonic type. Shifting foci of activity can be made out, with later concentric bodies cutting earlier ones. In Mull two foci can be made out, whilst in Ardnamurchan there are three (fig. 83), the most easterly being the earliest. Walker (1975) has recently discussed the evolution of the British intrusive centres. He sees an early event in the history of these centres as the uprise to a high level of an acid diapir. Once a body of acid magma has been generated and its rise as an acid diapir initiated, events then follow a set pattern and the development of the centre proceeds automatically for so long as magmatism continues to send up batches of basaltic magma. Walker believes that the emplacement of cone sheets is governed by the tendency for rising magma to move in the direction of maximum excess hydrostatic pressure. Walker's evolutionary cycle accounts for the granite plutons, cone sheets, ring dykes and the layered intrusions (the Cuillins). Moreover it takes into account the great gravity "highs" thought to be cylinders of basic intrusives which aid the rise of the acid magma. Fig. 84 shows the various stages in Dr. Walker's sequence for these centres.

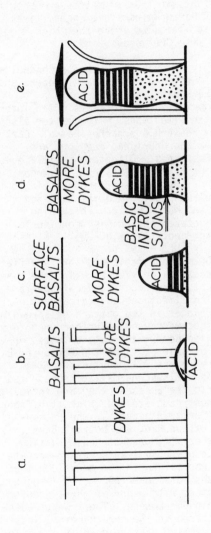

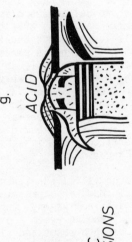

Fig. 84. Evolution of a Tertiary igneous complex (after Walker, 1975).

The dyke swarms trend mainly NW-SE. Extensive tension must be invoked to
account for the great overall width covered by the dyke material, mostly
olivine dolerite or teschenite. They must have fed flows long since removed
by erosion. The dykes extend as far south as Lundy, Dingle, Snowdonia and
the N.W. Midlands. In Snowdonia and Anglesey, the dykes probably belong to
the Mourne swarm and George (1974) believes "it is possible, perhaps
probable, that a basaltic cover was formerly a blanket over parts of Gwynedd".
The Eocene dykes on Snowdonia occur at heights of 750-780 metres and show that
there has been a good deal of unroofing of North Wales since Eocene times. As
George states, the present land surface, at altitudes above the "high
plateau", is not of Eocene, still less of late Cretaceous, age.

Palaeocene and Eocene sediments occur thinly through St. George's Channel and
into the Celtic Sea with chalk deposition continued into the Palaeocene and
with carbonates and calcareous clays in the Eocene. In the Western
Approaches, Eocene sediments are shallow marine sands and clays often interbedded with deltaic sediments and freshwater limestones. The latter also
characterise the Oligocene in parts of the Western Approaches Basin. Early
Tertiary sediments are possibly absent over the Hebridean area but are
developed to the west of the Orkneys and Shetlands. Oligocene argillaceous
sediments occur however in the Sea of the Hebrides. Tertiary limestones and
clays occur on both sides of Colonsay. Tertiary sediments occur again in the
West Shetland Basin. Thin Tertiary probably connects into the Sule-Sgeir
and North Minch basins in the vicinity of the Minch Fault. Thickening
Tertiary sediments characterise the basin between the British continental
shelf and the Faeroes Shelf (see Naylor and Mounteney, fig. 28).

The Rockall trough is filled with over 3000 m of Mesozoic and Tertiary flat-
lying sediments. How far back into the Mesozoic (or even Permian) the basal
sediments extend is not known. An important break occurs under (Upper?)
Eocene sediments, probably semi-consolidated limestones and calcareous oozes
passing eastwards into sands and silts. Changes in the pattern of spreading
in the North Atlantic during the early Oligocene caused a further unconformity followed by renewed Neogene sedimentation comprising further oozes or
semi-consolidated chalk. Thin Tertiaries overlie parts of the Porcupine Bank
(west of Ireland) but much thicker Tertiary covers fill the intervening
trough known as the Porcupine Seabight. The Palaeocene may be represented by
chalky sediments and these could extend into the Eocene. They are followed
by Neogene silts and clays. Perhaps the most complete and thickest offshore
Tertiary sequence in the sea areas west of Britain occurs in the Hatton-
Rockall Basin (on Rockall Plateau between Hatton Bank and Rockall Bank). In
this basin, probably another abortive line of attempted spreading dating
back to Mesozoic times, Palaeocene to Pleistocene sediments were penetrated
in two drill holes located in this Hatton-Rockall Basin (in over 1000 m of
water). Neogene chalks and calcareous oozes total over 800 m. Palaeogene
sediments are frequently of chalk or limestone with bands of chert nodules.
Breaks occur within the Oligocene and within the Eocene. The lowest Eocene
and underlying Palaeocene were only encountered in the more marginal-located
of the two holes, and were graded sands and clays passing down into pebble
beds.

Crustal instability in Western Britain outlasted the Eocene igneous activity
and may again be related to changes in spreading pattern and/or spreading
rates in the North Atlantic. As Brooks has shown, the same N-S belt of
Western Britain was again affected with block-faulting localising a series of

fault-controlled Oligocene basins filled with thick lacustrine sequences.
This sequence is 350 m thick in the Lough Neagh Basin of N.E. Ireland and
600 m thick in the Tremadoc Bay Basin of Cardigan Bay. A new basin has been
located east of Lundy Island, a basin whose western edge is sharply cut off
by the Sticklepath Fault. Boreholes into the basin proved silts, mudstones,
lignitic clays and lignites of middle to upper Oligocene age (Fletcher, 1975).
Brooks and James (1975) suggest that the sediments may total almost 350 m in
thickness. A small patch of Oligocene occurs near the same fracture at
Flimston in South Pembrokeshire. In Devon, Oligocene occurs in the Petrockstow and Bovey Tracey basins. Over 600 m of Oligocene occurs in the former
locality whilst the thickness in the latter may exceed 1000 m. Freshney
(1970) has suggested that the Petrockstow material could be linked with a
river system flowing along the fault zone into the sea in the Bristol Channel
area.

The marked Palaeogene crustal and magmatic unrest in Western Britain must
relate to North Atlantic sea-floor spreading (successful and abortive) in
areas west of Britain but with the overall accent on net uplift rather than
subsidence. At the same time, net sagging was occurring in the south-eastern
part of Britain and in the North Sea. Maybe eastward movement of the British
area (linked with Atlantic spreading) was being impeded somewhere east of the
North Sea thereby causing the compressive sag. Whatever the cause, the
Palaeogene Sea invaded the southeastern parts of England (fig. 85b). The
invasion was mainly from the North Sea direction and at its maximum the
Palaeogene Sea covered the North Sea, Denmark, Northern Germany, the Low
Countries, the Paris Basin and S.E. England. From the latter Anglo-French
areas, marine connections existed at times with the Atlantic via the Western
English Channel (allowing Mediterranean foraminifera to migrate in). Boreal
influences via the northern North Sea are suggested by certain molluscan
faunas.

The greatest subsidence took place in the central portions of the North Sea
where over 2000 m of Tertiary sediments (3500 m in the Ekofisk region,
2700 m in the vicinity of Auk) are monotonously shaley with minor bands of
carbonate in the Palaeogene. The shales are occasionally multicoloured or
red. A distinctive top-Palaeocene marker is a 30 m unit of volcanic tuffs
(54 m.y. old) with sandy bands. Palaeogene thicknesses total amost 1000 m
for Argyll Field and nearly 1300 m for Forties Field. The Palaeogene
sediments thin both towards the extreme north and extreme south-west of the
North Sea area and become more sandy in the latter direction).

Palaeogene sediments are best seen in S.E. England, in two areas - the London
Basin and the Hampshire Basin (the latter including the northern portion of
the Isle of Wight). In the London Basin, Palaeocene and Eocene beds are
preserved (up to the Auversian Stage) but there are no Oligocene remnants. In
the Hampshire Basin, Lower and Middle Oligocene strata are also preserved. In
the Isle of Wight, the Palaeogene totals about 700 m, the Oligocene forming
the upper 200 m of this thickness. The Palaeogene Sea was subject to appreciable changes in area over S.E. England and periodically advanced and
retreated over that area. The greatest advance was at the time of deposition
of the London Clay, preserved today as far west as Newbury and as far north
as Yarmouth. The westward maximum extent and eastward retreat of this
Ypresian sea (depositing the London Clay) is shown in fig. 85b. The two S.E.
England Palaeogene basins have of course been isolated by the mid-Tertiary
folding. Originally, Palaeogene deposits spread across the whole of the

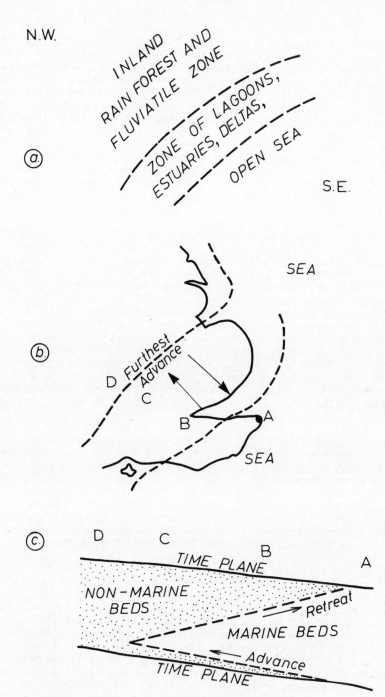

Fig. 85.  Eocene cycles of sedimentation (after various authors).

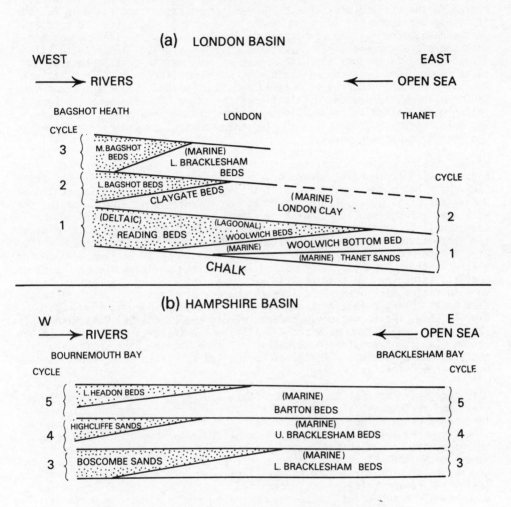

Fig. 86. Palaeocene and Eocene formations in S.E.
England (after Stamp and after Wells and Kirkaldy).

southeast.  It must be remembered that the marine advances spread in from
down the North Sea.  As a result Palaeogene successions are more dominantly
marine in the extreme east of S.E. England (Kent-Essex) and again in Belgium.
It is however important to remember that the periodic advances and retreats
of the Palaeogene sea can be matched in the Anglo-French-Belgian areas and
must therefore represent major eustatic changes, probably linked with Atlantic spreading variations and with consequent changes in areal growth of
ocean floor, land areas or volcanic ridges.  Important changes in N.Atlantic
spreading pattern occurred between 30 and 40 m.y. ago.  Important transform
faults occurred across the N.Atlantic at this time.  Marine advances were
rather more limited in these Oligocene times than in the preceding Eocene
and this fact may again be linked with the important spreading variations in
the nearby developing ocean.

Several cycles can be identified in these Palaeogene sediments of S.E.
England, each cycle being marked by a marine advance and a marine retreat.
Each cycle has a non-marine and a marine portion and these are understandably
wedge-like (fig. 85c).  It follows that the advances and retreats of the sea
will have transgressed the true time planes so that the marine and non-marine
wedges will be diachronous.  Also in fig. 85 it can be seen that at point A
the cycle could be entirely marine and at point D entirely non-marine while
at the intermediate localities B and C there would be an earlier advance and
a later retreat of the sea, producing intermediate environments or facies of
mudflat, estuarine, deltaic or lagoonal type.  Studies of the varying facies
show in fact a marked zonal pattern to the environments from open sea in the
east to an inland rain forest and fluviatile zone in the west (fig. 85a).
Fig. 86 shows the formational variants within the London and Hampshire basins,
(after Stamp and after Wells and Kirkaldy).  The swept-in flora found in the
London Clay is of tropical or sub-tropical type and the Nipa fruits are found
today on Indian coasts.  Remains of crocodiles have also been found, together
with over 500 species of plants.

Oligocene sediments in the Hampshire Basin are largely composed of marls,
clays and freshwater limestones (such as the Bembridge Limestone).  These
limestones contain freshwater gastropods such as Planorbis.  Marine incursions were probably short-lived, the thickest marine representative being
the uppermost Oligocene sediments (the Upper Hamstead Beds) preserved in the
Hampshire Basin.  Other sediments reflect lagoonal, estuarine and lacustrine
environments.  Cerithids such as Batillaria  occur today on Mexican mudflats
(Daley and Edwards, 1974).  Colour-banded micritic "crusts" on limestones
occur today on subaerially exposed, modern Caribbean limestone surfaces.  The
base of the Oligocene is generally placed at the base of the Middle Headon
Beds (Brockenhurst marine beds) but there is some evidence from the ostracod
and mammal faunas for placing it much higher in the sequence, within the
Bembridge Marls.  An interesting band in the latter is the "Insect Limestone"
containing the wings of fungus flies.  The mould of a fossil feather was
discovered on a recent excursion of the Geologists' Association.

No Miocene sediments are known on land in Britain and this was largely a time
of deformation involving folding (chiefly in the south) and faulting.  These
deformations were really the northern and outward pulses of the major Alpine
Storm which was throwing up the great Alpine mountain chains across southern
and central Europe, as the African and European plates approached one another.
It should be noted however that, as Ager has recently pointed out, "it is
important to remember that the coming together of Europe and Africa was a

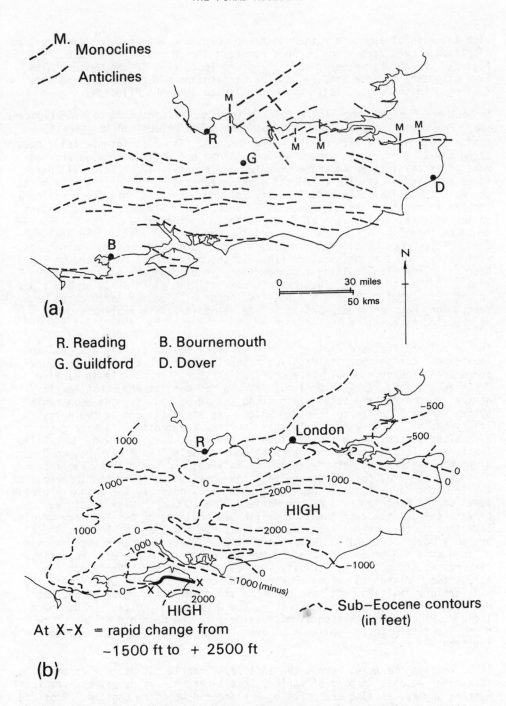

(a)

R. Reading    B. Bournemouth
G. Guildford  D. Dover

At X-X = rapid change from
       −1500 ft to  + 2500 ft

(b)

Fig. 87. Aspects of the Tertiary warping of South-East
England (after Wooldridge and Linton).

slow and bashful approach rather than a precipitate embrace". A varying time
of climax occurs across the various European Alpine chains (see Ager, 1975,
fig. 7). In the Pyrenees it was Palaeocene to Eocene, in the Western Alps
late Oligocene to late Miocene, in the Carpathians mid-Cretaceous and late
Miocene. In the Juras, later movements lasted into the Pliocene.

In Southern England, the Alpine folding involves sediments up to Mid-Oligocene
age. Presumed Pliocene planations cut across the folded earlier strata, so
that a broad Miocene deformation is reasonable. The most intense folds occur
along the Isle of Wight-Purbeck-Weymouth line, probably an echelon arrange-
ment of at least three separate structural elements. These folds all have
steep and (in places like Lulworth) even somewhat inverted northern limbs to
the main anticlinal elements. Gravity folding occurs on this steep northerly
limb in incompetent horizons such as Purbeck. The intensity of the warping
in the vicinity of the Isle of Wight is shown by the range of the sub-Eocene
contours (see fig. 87b). This figure also shows the "high" of the Wealden
Dome. This broad upfold, an "inversion" structure, embraces many minor east-
west folds. On its northwestern flank is the more marked Guildford or Hog's
Back anticline with faulting accompanying or replacing the steep northern
limb. In the vicinity of the lower Thames, folding along NE-SW or N-S lines
occurs (fig. 87a). These N-S monoclines occur also on the north coast of
Kent. They may represent warping in the Mesozoic-Tertiary blanket over
rejuvenated basement faults.

In Central and Northern England, a mid-Tertiary age is usually assigned to
faults that can be shown to displace Mesozoic strata, but it must be remembered that fracturing could have occurred earlier, for example in earlier
Cretaceous times. In S.W. England, Dearman has demonstrated that Tertiary
dextral movements have taken place along the NW-SE faults. One such fault
system is the Sticklepath fracture which cuts the Oligocene of the Bovey
Tracey basin. It is very likely that Tertiary movements could have occurred
also along several Welsh faults, such as the Bala, Vale of Clwyd, Moel Gilau
fractures. Tertiary movement, perhaps considerable, could have happened
along the eastern boundary fault of the Malverns and the Mochras Fault has
obviously suffered post-Oligocene faulting equal to at least the thickness of
the Tremadoc Bay Oligocene sequence (650 m) plus possibly the height above sea
level of the Cambrian mass of the Harlech Dome (800 m). Many fractures
beneath the sea floor of the South Irish Sea could be of mid-Tertiary origin.
Important north-Pennine fractures such as those bounding the Pennine Block
probably moved (outwardly) in Tertiary times. These movements were imposed
on to a main N-S broad upfold along the whole Pennine range with radiating
flexures at its southern extremity from Cheshire and North Staffordshire
across the Midlands to Leicestershire. In Scotland and the seas around it,
mid-Tertiary faulting must have occurred along the Minch, Camasunary, Outer
Hebrides, West Shetland and, probably, the Great Glen faults. In Northern
Ireland, Miocene deformation is most evident in the downfold of the Lough
Neagh Syncline with an amplitude of over 1000 m and in its accompanying
faulting.

It is tempting to think beyond the individual ripples of Miocene deformation
to a more regional warping of various British areas. Wales may be a regional
Miocene upwarp, as also are parts of Ireland and Northern England. Scotland
almost certainly is a major upwarp. These upwarps need not have been
symmetrical. Intervening areas such as the Irish seas were correspondingly
major downwarps. How much a part was played in this regional contrast of

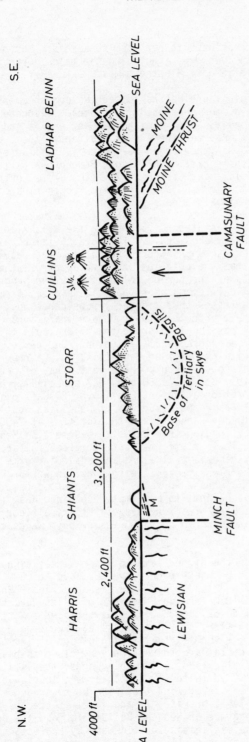

Fig. 88. Projected summit profiles in Hebridean Scotland (after George, 1966).

broad structure by boundary fractures is not easy to assess.  Moreover the
rising swells and intervening downwarps (including the major North Sea downwarp) were to continue active in post Miocene times.

With this broad regional warping, one begins to see the commencement of the
final phases of the moulding of the present British area.  Pliocene sediments
are very restricted on land today, being limited to East Anglia and to
pockets near Bideford and Camborne (on the 400 ft platform), the Lenham Beds
of the North Downs (above 180 m) (they could be Miocene), similar sands and
gravels on the South Downs, the pocket at Flimston, Pembrokeshire (probably
preserved in a Miocene downfold - see George, 1974, fig. 94).  Further
"Pliocene" gravel pockets occur in N.E. Wales (Halkyn Mountain) and on the
Pennines.  The North Wales residuals appear to be preserved because they
collapsed into the Carboniferous floor (fig. 89b).  They may even be earlier
than Pliocene.  As George (1974) states, "evidence in the Welsh landscape of
events in Neogene times is mainly erosional".  The word "British" might well
be substituted for the word "Welsh".  As George suggests, "the resulting
(Neogene) landforms suggest that the principal erosive agent was the sea, and
that at least the later events of the period followed from the progressive
emergence of Wales, as more widely of much of the British area, from beneath
sea level".  George rightly stresses that a general subsidence must have been
antecedent to the late Neogene emergence.  The stubborn traditional faith in
a long-persisting Chalk cover has for long prevented clear, bold thinking on
the latter stages of Britain's topographical evolution.  One has only to
stand on any high point in Britain, be it on Ben Nevis, the Cairngorms, Great
Gable, Snowdon or the Cuillins to see around one a vast array of denuded
platforms at different heights.  George (1966 and 1967 especially) has
clearly demonstrated the truncation at heights well over 2000 ft of platforms
across the complex geological frame and structure of Hebridean Scotland and
Northern Ireland.  Fig. 88 clearly shows this with platforms at 2400 ft and
at 3200 ft bevelling a geology ranging from Lewisian basement to Mesozoics
and folded Tertiary basalts.  These platforms must represent pulsatory uplifts
through a considerable height range.  In Ireland and again in North Wales,
high level platforms are to be clearly seen. George, Hollingworth and Brown
have stressed the wide extent of these surfaces.  They are too persistent,
too widespread, too uniform in height, to be simply explained away (as O. T.
Jones suggested) by simple warping of a sub-Mesozoic floor, a floor at well
over 2500 ft in North Wales and down to only 200 ft in South Wales coastal
platforms.

The teasing problem is whether the higher-level platforms, say higher than
800 ft O.D., are marine-planed or whether only the 200, 400 and 600 ft platforms are of marine bevelling and the higher ones are subaerial.  However,
even if platforms such as the 1800 ft High Peneplain of Wales, or the
2400 ft surface of West Scotland or the 3000 ft plus surface of Northern
Scotland are of subaerial planation, they must still have undergone sufficiently long planation to make them virtually old-age (or at least very
mature) stages of fluviatile cycles and thereby be very close to base levels
at their very mature stage of planations.  In other words, they still suggest
very much lower positions relative to their particular sea levels whether the
surfaces were above or below sea level when they were being eroded.  A
considerable net fall of sea level or rise of land level is thereby to be
inferred and a very major emergence of Neogene Britain to form the uplands of
today.  The writer prefers the explanation put forward by Professor George
(1974).  As George indicates, "both gentle folding ... and hinterland uplift

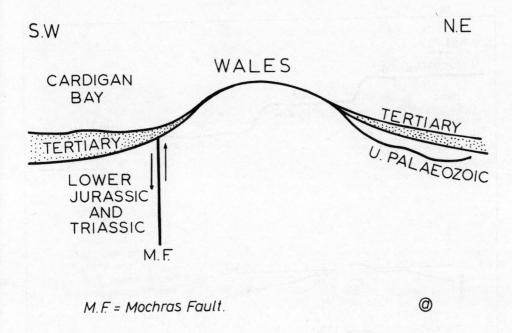

S.W                                                      N.E

CARDIGAN
BAY

WALES

TERTIARY

U. PALAEOZOIC

TERTIARY

LOWER
JURASSIC
AND
TRIASSIC

M.F.

*M.F. = Mochras Fault.*                                 ⓐ

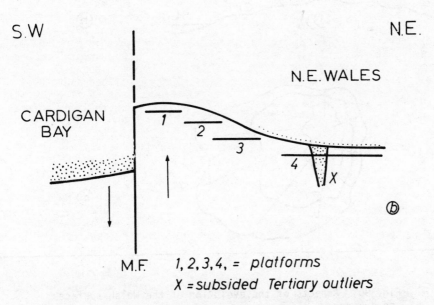

S.W                                                     N.E.

N.E. WALES

CARDIGAN
BAY

1
2
3
4
X

M.F.                                                    ⓑ

M.F.   *1, 2, 3, 4, = platforms*
       *X = subsided Tertiary outliers*

Fig. 89.   The Tertiary evolution of Wales (in part, after
Walsh and Brown, 1971)

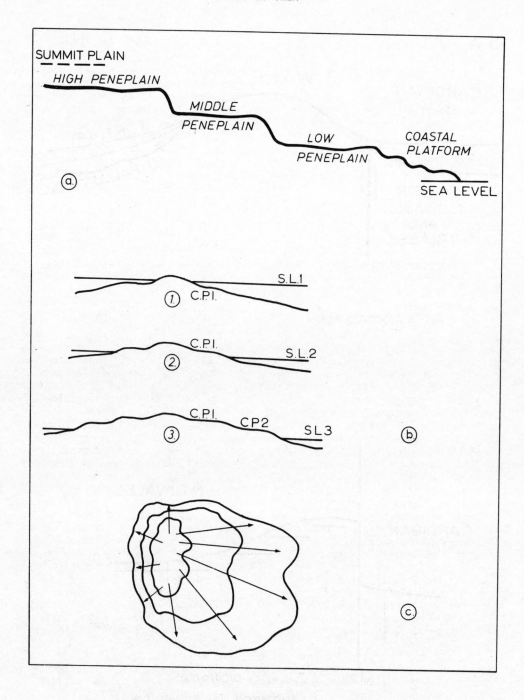

Fig. 90.  Aspects of the evolution of the Welsh surface
and drainage.
(a)  Welsh high-level platforms (after Brown, 1960)
(b)  Suggested evolution of these platforms
(c)  Ancestral drainage pattern for Wales
C.P.1.  Coastal Platform 1.  S.L.1. Sea Level 1.

were instrumental in determining the late-Neogene landforms: the exposed
land surface of Wales and southern England may then be looked on as mild
swells rising out of the Pliocene sea in complement to the synclinal sags,
long-sustained, of the southern Irish Sea and the North Sea". The Pliocene
sea notched the emergent swells as they rose, their margins cliffed and
retreating at each stage of uplift. A pulsed uplift, with intervals of still-
stand, encouraged the erosion of relatively wide wave-cut benches to form a
stepped profile, the step at any one stage being insignificantly warped.
The growing swells may have had then the form of exceedingly gentle anticlines but the domed surface would in its profile become dominated by the
marine benches. Any one wave-cut bench sloped seawards from its coastal
cliff at an angle often diverging little from the slope of the gradually
emerging dome. Fig. 90a shows Brown's main platforms for Wales. Fig. 90b
shows the slow persistent doming of a Wales out of the Pliocene sea which
notches wide levels at each time of "still-stand". As a result, a supposed
eustatic variation need not be eustasy at all but rather a very gentle, but
very persistent, doming or "swelling" of what are the upland areas of today.
It follows that Scotland may have been the area where doming reached the
greatest net amplitude as this is where the notches reach their greatest
height.


## THE DRAINAGE PATTERN

The other major problem of Neogene geomorphology is the evolution of the
British drainage patterns. Much has been written on the evolution of regional
(and local) drainage patterns. In many accounts accent has been placed on
the superimposed character of the drainage. More simple covers have been
involved to explain the way in which rivers now cross very complex geological
patterns with often only very local or minor secondary adaptations to geology
or structure. Of these simple intervening covers, the most popular has been
the Chalk, invoked, for example, to account for the southward or southeastward
directed drainage of South Wales (across the hard rims of the Coalfield) or
for the radial drainage of the Lake District. Two things have been forgotten
in this supposed superimposition of drainage on Chalk. Firstly there is now
growing evidence to show that the chalk cover was deformed, fractured and
removed over very wide parts of Britain long before the ancestral British
drainage patterns began to form. Even if these first rivers are taken back as
far as the Miocene period, then post-Cretaceous to pre-Eocene and post-Eocene
to pre-Oligocene movements, shown by George to have been widespread and
effective in Ireland and West Scotland, would by then have removed any remnant
Chalk cover (and probably much pre-Chalk Mesozoic as well). If there is no
Chalk in the Bristol Channel and Cardigan Bay (and there isn't any!) - the
areas of downfolding - then it's very unlikely that a Chalk cover would have
survived for any considerable length of time into the Tertiary. Again there
is no Chalk between Oligocene and Lias in the Mochras Borehole. The Chalk
of Mochras (if it was there in the first place) was removed before Oligocene
deposition commenced.

The second thing that has been forgotten is that rivers can be directly
superimposed on to surfaces - even those of complex geology as in Northern
Ireland (see George, 1967). If there's a slope a river will flow directly
down it. With successively emerging surfaces ("swell") out of a Pliocene
sea, the earlier restricted drainage will flow outwards on to the extending
area each time flowing across the margins from higher levels to lower ones

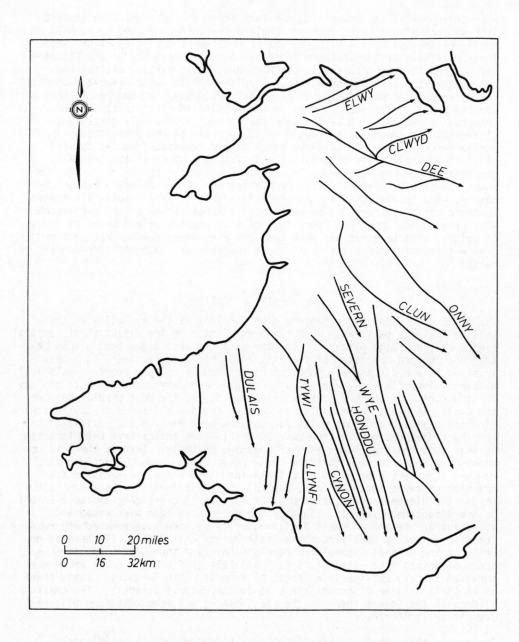

Fig. 91. Ancestral Welsh drainage pattern (after Brown, 1960).

(from Coastal Plateau 1, or CP1, on to CP2 in fig. 90b. If the emerging
areas had  asymmetrical levelled margins, as in fig. 90c, then rivers could
be longer in some directions than in others. This may have happened in the
case of an emerging Wales. Bevels would have continuously cut deep into the
west side (and north side) of Wales because of the softer Mesozoics adhering
to these sides of the Welsh "oldland".  Platforms would be narrow on these
sides and almost non-existent over some stretches. (The north coast of
Exmoor might be a similar situation, with as a result a "hog's back" drop
from an appreciable height virtually down to sea level). Coastal faults may
even have been contemporaneously moving, thereby accentuating the same limit
of bevel along such coasts. As a result of the asymmetrical "peak-line" of
emergent Wales, the Welsh ancestral river pattern had a long eastern, south-
eastern and southern element (fig. 91). Ancestral rivers such as the Conway-
Dee flowed eastwards (Linton believed it became the Trent). The Onny, Severn
and Wye flowed southeastwards, possibly towards a Thames catchment. Rivers
such as the Loughor, Towy, Dulais flowed southwards. Brown (1960) has traced
the evolution of this ancestral  asymmetrical pattern through the "High
Plateau", "Middle Peneplain" and "Low Peneplain" stages. By the latter, major
adaptations to structure (along the Bala Fault, Teifi Anticlinorium, Swansea
Valley Fault at Glasbury, on the Wye) had begun to occur. Further adaptations
(Neath, Tawe, Ystwyth, Lower Towy etc.) occurred with the commencement of
emergence of the coastal platforms.

Similar explanations may serve for other British areas.  Asymmetrical bevel
patterns probably apply also for Northern Scotland. A symmetrical pattern
must have occurred over the Lake District area. The axis of the Pennines
emergence had a N-S orientation and rivers flowed to the east and west. The
evolution of the Weald drainage has been known for some time. In this case,
a chalk cover has obviously played an important part.


## THE PLEISTOCENE ICE AGE

The final chapter in the story of the British area belongs to the Pleistocene,
a time of increasing cold at first and then of alternations of cold and less
cold conditions thereafter. Now the latest chapter (not yet even complete)
ought to be more clearly known and understood than any that preceded it in
the geological story, and yet this Pleistocene story is the most controversial
of them all with varying interpretations of the British Pleistocene sequences
and events. The last 40,000 years of time is now being more uniformly and
more orderly described because of the use of C14 dates, but agreement is still
lacking regarding the story before this. On the other hand, the challenge
presented by the evidence has been a most stimulating one. The task of
reconstruction is an immense one. Pleistocene deposits are, by their very
origin, tricky deposits to read and interpret. They are mostly land deposits,
subject to all the difficulties of reconstructing continental-type deposition.
Many have been moved to their present place by ice sheets or glaciers and
many may therefore have been moved yet again by later advances of ice or redistributed by meltwaters. Seafloor deposits or shells may have been ripped
up by ice and dumped long distances away, on mountain slopes in Snowdonia for
example. Fossil evidence must be treated carefully again because of carriage
from its original resting place by ice or by water. A Pleistocene succession
must be read with great caution and with constant regard to successions in
that area. In Gower, South Wales, Bowen has shown that some boulder clays in
certain coastal sequences were not deposited just there and represent
redeposited glacial drift.

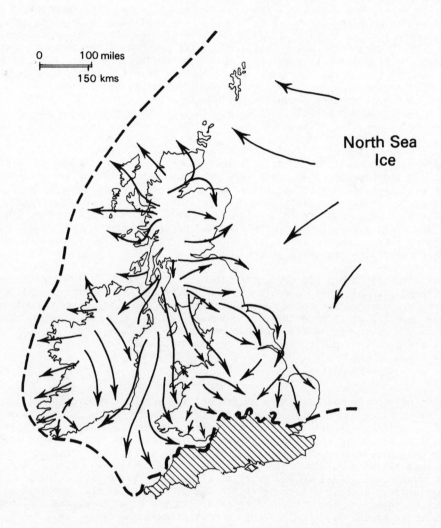

Fig. 92.  Maximum ice extent in the British Isles (after several authors).

The task of reconstruction has then been a difficult one and yet it has had
its rewards. Wonderful new tools have been discovered such as the use of
pollen analysis, stone orientations, pattern ground and beetle faunas.   The
latter, studied especially by Dr. Coope, have been shown to have rapid shifts
of population in response to the changing climate.  Palynology also reveals
climatic changes.  Pollen is widely produced and moreover is widely dispersed
finding its resting place in all sorts of environments and deposits.  Pollen,
like beetle carapaces, are very resistant to physical and chemical changes.
Pollen diagrams make correlation of sequences possible.

One confusing feature of Pleistocene reconstruction and stratigraphy has been
the constant changing of nomenclature.  Four major periods of cold (or near
cold in one instance, the first) have generally been identified in Britain.
These four have been referred to the nomenclature in the Alps (Gunz, Mindel,
Riss and Wurm) or in Scandinavia.  Other schemes used such terms as "the
Great Eastern Glaciation" or "the Lower Chalky Boulder Clay" or "Older Drift"
and "Newer Drift".  More recently nomenclature using type localities has been
fashionable.  The second cold period (if we think of the East Anglian Crag
deposits as representing the first somewhat colder time) has been called the
Lowestoftian and the third the Gipping or Gippingian.  The most recent recom-
mendation is that of the Geological Society (Mitchell, Penny, Shotton and
West, 1973) and is as follows:-

|  | Flandrian | Type site |
|---|---|---|
| 4. | Devensian | Four Ashes, Staffs |
| I. | Ipswichian | Bobbitshole, Ipswich |
| 3. | Wolstonian | Wolston, Warwicks. |
| I. | Hoxnian | Hoxne, Suffolk |
| 2. | Anglian | Corton |
| I. | Cromerian | West Runton |
| 1. | Beestonian and earlier stages | |

The numbers refer to the four major colder periods (the first was probably
complex and less cold than the other three).  The milder intervening interglacial periods are marked I.  The Devensian was complex (perhaps because
more detail is known about it) with intervening advances (stadials) and
retreats (interstadials) within it.  Certain minor "readvances" of ice took
place late in the Devensian, more particularly in Scotland (Highland Readvance, etc.).  The terms Devensian, etc. are referred to as stages though
obviously they do not compare in length of time with stages in older
geological divisions.  The base of the Devensian can be considered to go back
to about 70,000 years B.P. (see Bowen, 1974, Table 14) with the bases of the
Middle Devensian, Upper Devensian and Flandrian drawn at 50,000,  26,000 and
10,000 years B.P. respectively.  Dating for the pre-Devensian divisions is
not really possible.  Pre-Devensian Pleistocene time spans back from 70,000
years B.P. to almost two million years ago.  The middle interglacial may have
been the longest of the three interglacials.  At its maximum extent (probably
in the Wolstonian or Gipping glacial phase) ice reached virtually to the tip
of S.W. England, the eastern end of the Bristol Channel and almost to the
London area (fig. 92).  If the theory of Dr. Kellaway is accepted then the
ice reached Salisbury Plain bringing the "bluestones" of the Prescelly Hills
as erratics to within reach of Stonehenge.

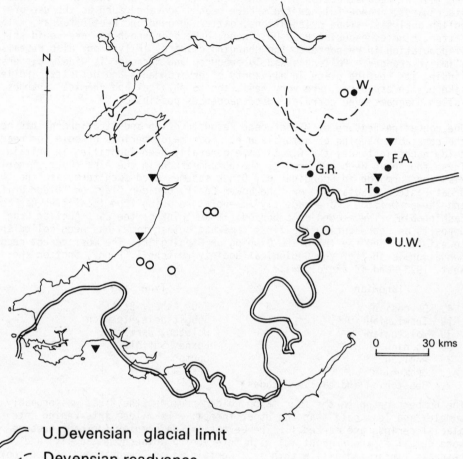

N

W

F.A.

G.R.

T

O

U.W.

0        30 kms

˷  U.Devensian  glacial limit

- - -  Devensian readvance

▼  Ice−wedge polygonal pattern ground

o  Pingos

●  Localities:−  W  Wybunbury      G.R. Great Ryton
               FA.  Four Ashes      T.  Tryssul
               O.   Orleton         U.W. Upton Warren

Fig. 93.  Glacial features in Wales (after Bowen, 1974).

Lower Pleistocene deposits occur almost exclusively in East Anglia, exceptions being the possibly Cromerian St. Erth Beds of Cornwall and the (Red Crag?) Netley Heath Beds, near Guildford. The Lower Pleistocene Crags of East Anglia extend from near Harwick in the south to Cromer in the north. The Red Crag's fauna shows a marked increase in cold or boreal species as compared with the Mediterranean-type faunas of the Pliocene Coralline Crag. Marked fluctuations of climate occurred during the deposition of the Red, Norwich and Weybourne Crags, as revealed by pollen and microfauna studies. The pollen indicate cold times of sparse Arctic heaths with intervening temperature phases. If Scandinavian ice was on its way it had not as yet reached Eastern Britain, except possibly for the odd ice flow during Weybourne Crag times.

The first main interglacial, the Cromerian, is best exemplified by the Cromer Forest Bed, estuarine sands and clays with rootlet beds and uprooted tree stumps. Pollen indications are of, at least at times, a relatively warm interglacial. This interglacial was centred about a time of probably 1.5 m.y. ago. It was after this warm episode that the real onset of cold conditions began to firmly take place. As Bennison and Wright (1969) point out, the climate must have become colder by at least 4°C, on average, throughout the year. Snows of winter were not dispersed during cooler summers. It is possible that during winter months high pressure predominated north of Britain giving easterly winds and snowfalls as depressions passed through on more southerly tracks. The second glacial stage, the Anglian (or Lowestoft) was again composite with perhaps two much colder portions. The first was mainly the time of Scandinavian ice invasion, bringing boulder clays with Scandinavian erratics to East Scotland and N.E. England coastal districts. A less cold phase is represented by the Corton Sands which in East Anglia separate a Cromer Till from the Lowestoft Till (the Lower Chalky Boulder Clay). The ice ("Great Eastern Glacier") moved into E.Anglia from the north-west and lobes went down into Oxfordshire (Northern Drift) and the northern edge of the London Basin. Clays in the Midlands and higher river terraces of the Thames belong to this Anglian or Lowestoft phase.

The Hoxnian (or Great) Interglacial was of considerable length and thereby complex with at least two cold portions within it. Deposits at Hoxne, Nechells and Clacton belong to this stage. High sea levels probably characterised some horizons within this long interglacial. The Boyn Hill (100 ft) terrace was of this time. So also were the Swanscombe gravels. Swanscombe Man belongs here (the skull probably belonged to the opposite sex). At Hoxne, the lake-side vegetation was of thick mixed forest with plenty of alder. Some deforestation by man is indicated. Late Acheulian hand axes, delicate flake-tools and at least one blade have been found. Many species of beetle have been identified by Dr. Coope. The Swanscombe gravels yielded, besides the famous skull, a mammalian fauna including straight tusked elephant, horse, giant ox, giant deer, Rhinoceros, pig, wolf, cave bear and hare (Wymer, 1974). Clactonian implements come from all levels in the gravel.

The Wolstonian or Gipping glaciation brought Chalky Boulder Clay to the Midlands and the Gipping Till to East Anglia. This was the time of the maximum movement of ice down the Irish Sea moving across Pembrokeshire into the Bristol Channel and down into the Celtic Sea. The Fremington tills of Barnstaple Bay are Wolstonian. Overlying raised beach deposits are Ipswichian (Kidson and Wood, 1974). The oldest terraces of the Severn and Avon belong

to the Wolstonian.  The Greater Highland Glaciation in Scotland was also of
this time.  "Lake Harrison", an ice-dammed lake in the Midlands,had its
maximum extent at this time, extending from Leicester to beyond Stratford and
almost to Moreton (see Shotton 1953, Bennison and Wright, 1969, fig. 16.7).
The Ipswichian interglacial is represented by pollen-bearing clays channelled
into the Gipping Till near Ipswich and by many river terraces (Severn, Avon,
Trafalgar Square).  In South Wales a 30 ft raised beach (Patella Beach) is
widespread (resting on a platform which could be older).  A still higher rise
of sea level is indicated by the Neritoides Beach of Gower (e.g. Minchin
Hole).  Organic silts at Four Ashes and Tryssul in the West Midlands (fig.93)
belong to the Ipswichian.

The Devensian ice sheets did not extend as far south as those of the Anglian
and Wolstonian.  The Upper Devensian glacial limit in Wales (after Bowen,
1974) is shown in fig. 93.  This limit ran across Central Gower in South
Wales and outside it angular head deposits represent preglacial conditions of
Devensian times.  At Pencoed, near Bridgend, (Devensian) Welsh Till rests on
(Wolstonian) Irish Sea Till (see Bowen, 1974, fig. 99).  In Eastern England,
lobes of Devensian ice reached York and Hunstanton.  Gradual retreats of the
Devensian ice-front caused many lakes to form (the Vale of Pickering Lake and
Lake Lapworth, the latter spilling to cut the Ironbridge gorge).  Several
late readvances took place, in Northern England and in Scotland.  Raised
beaches at 100 ft and 50 ft correlate with the Perth Readvance and Highland
Readvance.

Flandrian times (post-glacial) began about 10,000 years B.C.  It was a time
of recovery over Scotland with a falling succession of raised beaches.  At
the same time a rise in sea level is evident in the south of Britain, with
successive rises in South Wales, for example, adding up to a net rise of sea
level of some 75 ft.  Several peat beds were recognised in foundation work
around the Swansea, Llanelli, Port Talbot and Cardiff coastal areas.  The
climax of the Flandrian transgression came with the formation of the Straits
of Dover and the drowning of the southern North Sea.  The British Isles
finally became isolated from the European mainland.  It has taken the Treaty
of Rome to join us together again.


## Suggested Further Reading

Ager, D.V. 1975.  The geological evolution of Europe. Proc. Geol. Ass. Lond.
   86, 127.
Bennison, G.M. & Wright, A.E. 1969.  The Geological History of the British
   Isles.  Edward Arnold, London.
Bowen, D.Q. 1974.  The Quaternary of Wales.  In: The Upper Palaeozoic and
   Post-Palaeozoic Rocks of Wales, (Ed. T. R. Owen).  Univ. of Wales Press,
   Cardiff.
Brooks, M. 1973.  Some aspects of the Palaeogene Evolution of Western
   Britain in the context of an underlying mantle hot spot.  Journ. Geol.
   81, 81.
Brown, E.H. 1960.  The relief and drainage of Wales.  Univ. of Wales Press,
   Cardiff.
Fletcher, B.N. 1975.  A new Tertiary basin east of Lundy Island.
   Jl. geol. Soc. Lond. 131, 223.
George, T.N. 1965.  The geological growth of Scotland.  In:  The Geology
   of Scotland, (Ed. G. Y. Craig).  Oliver and Boyd, Edinburgh and London.

George, T.N. 1966. Geomorphic evolution in Hebridean Scotland.
  Scott. J. Geol. 2, 1.
George, T.N. 1967. Landform and structure in Ulster. Scott. J. Geol.
  3, 413.
George, T.N. 1974. Prologue to a geomorphology of Britain. Inst. British
  Geographes. Special Publ. No. 7, 113.
Kent, Sir Peter, 1975. Review of North Sea Basin development.
  Jl. geol. Soc. Lond. 131, 435.
Mitchell, G.F., Penny, L.F., Shotton, F.W. & West, R.G. 1973.
  A Correlation of Quaternary Deposits in the British Isles.
  Geol. Soc. Lond. Special Report No. 4, 1.
Naylor, D. & Mounteney, S.N. 1975. Geology of the North-West European
  Continental Shelf. Vol. 1. Graham Trotman Dudley Publishers, Ltd. London.
Rayner, D.H. 1967. The Stratigraphy of the British Isles. Cambridge
  University Press, Cambridge.
Steel, R.J. & Wilson, A.C. 1975. Sedimentation and tectonism (?Permo-
  Triassic) on the margin of the North Minch Basin, Lewis.
  Jl. geol. Soc. Lond. 131, 183.
Walker, G.P.L. 1975. A new concept of the evolution of the British Tertiary
  intrusive centres. Jl. geol. Soc. Lond. 131, 121.
Walsh, P.T. & Brown, E.H. 1971. Solution subsidence outliers containing
  probable Tertiary sediment in north-east Wales. Geol. Journ. 7, 299.
Wood, A. & Woodland, A.W. 1968. Borehole at Mochras, west of Llanbedr,
  Merionethshire. Nature, 219, 1352.
Woodland, A.W. (ed.). 1971. The Llanbedr (Mochras Farm) borehole.
  Rep. Inst. geol. Sci. 71/18.

# Additional Bibliography

## Chapter 1

Blackett, P.M.S. 1961. Comparison of ancient climates with ancient latitudes deduced from rock magnetic measurements. Proc. R. Soc. A. 263, 1.

Blackett, P.M.S., Bullard, E.C. & Runcorn, S.K. et al. 1965. A symposium on continental drift. Phil. Trans. R. Soc. A. 258, 1.

Condie, K.C. 1976. Plate Tectonics and Crustal Evolution. Pergamon Press, Oxford.

Falcon, N.L. & Kent, P.E. 1960. Geological results of petroleum exploration in Britain 1945-1957. Mem. Geol. Soc. Lond. No. 2.

George, T.N. 1962. Tectonics and Palaeogeography in Southern England. Sci. Progr. Lond. 50, 192.

George, T.N. 1963. Tectonics and Palaeogeography in Northern England. Sci. Progr. Lond. 51, 32.

Runcorn, S.K. 1962. Palaeomagnetic evidence for continental drift and its geophysical cause. In "Continental Drift" (ed. Runcorn, S.K.). New York Acad. Press.

Seyfart, C.K. & Sirkin, L.A. 1973. Earth History and Plate Tectonics. Harper and Row, New York.

Smith, A.G. & Briden, J.C. 1977. Mesozoic and Cenozoic Palaeocontinental Maps. Cambridge Univ. Press, Cambridge.

Tegrum, R.M. & Rees, G. 1975. Geology of the North West European Continental Shelf. Vol. 2. The North Sea. Graham Trotman Dudley Publishers Ltd., London.

Vine, F.J. & Matthews, D.H. 1963. Magnetic anomalies over oceanic ridges. Nature, London. 199, 947.

Wyllie, P.J. 1976. The Way the Earth Works. Wiley & Sons Inc., New York, London.

## Chapter 2

Bassett, D.A. & Walton, E.K. 1960. The Hell's Mouth Grits: Cambrian greywackes in St. Tudwal's peninsula, North Wales. Quart. Journ. Geol. Soc. Lond. 116, 85.

Bloxam, T.W. & Allen, J.B. 1960. Glaucophane-Schist, Eclogite and associated rocks from Knockormal in the Girvan-Ballantrae Complex, South Ayrshire. Trans. Roy. Soc. Edin. 64 (No. 1), 1.

Cowie, J.W. 1974. The Cambrian of Spitzbergen and Scotland. In: Lower Palaeozoic rocks of the World, Vol. 2. Cambrian of the British Isles, Norden and Spitzbergen. London.

Cowie, J.W. & Rushton, A.W.A. 1974. Palaeogeography of the Cambrian of the British Isles. In: Lower Paleozoic rocks of the World. Vol. 2. Cambrian of the British Isles, Norden and Spitzbergen. London.

Dewey, J.F. 1963. The Lower Palaeozoic stratigraphy of central Murrisk, County Mayo, Ireland, and the evolution of the South Mayo trough. Quart. Journ. Geol. Soc. Lond. 119, 313.

Greenly, E. 1919. Geology of Anglesey. Mem. Geol. Surv. Great Britain, 2 vols.

Harland, W.B. & Rudwick, M.J.S. 1964. The great Infra-Cambrian Ice Age. Sci. Am. 211, 28.

Hughes, N.F. (ed.) 1973. Organisms and continents through time. Spec. Pap. Pal. Ass. 12.

Johnson, M.R.W. & Stewart, F.H. 1963. The British Caledonides. Oliver and Boyd, Edinburgh and London.

Johnson, M.R.W. 1965. Torridonian and Moinian. In: The Geology of Scotland, (Ed. G.Y. Craig). Oliver and Boyd, Edinburgh and London.

Johnson, M.R.W. 1965. Dalradian. In: The Geology of Scotland, (Ed. G. Y. Craig). Oliver and Boyd, Edinburgh and London.

McKerrow, W.S. 1962. The chronology of Caledonian folding in the British Isles. Proc. natn. Acad. Sci. U.S.A. 48, 1905.

Moseley, F. 1964. The succession and structure of the Borrowdale Volcanic rocks north-east of Ullswater. Geol. Journ. 4, 127.

Read, H.H. & Watson, J. 1975. Introduction to Geology, Vol. 2. Earth History. Part II. Later Stages of Earth History. Macmillan Press Ltd., London and Basingstoke.

Roberts, J.L. 1966. Sedimentary affiliations and stratigraphic correlation of the Dalradian rocks in the South-west Highlands of Scotland. Scott. J. Geol. 2, 200.

Sadler, P.M. 1974. Trilobites from the Gorran Quartzites, Ordovician of south Cornwall. Palaeontology 17, 71.

Shackleton, R.M. 1969. The Pre-Cambrian of North Wales. In: The Pre-Cambrian and Lower Palaeozoic Rocks of Wales (Ed. A. Wood). Univ. of Wales Press, Cardiff.

Stubblefield, C.J. 1956. Cambrian palaeogeography in Britain. Int. Geol. Congr. (20th, Mexico). Sect. 1, 1.

Sutton, J. & Watson, J. 1970. The Alderney Sandstone in relation to the ending of plutonism in the Channel Islands. Proc. Geol. Ass. Lond. 81, 577.

Walton, E.K. 1963. Sedimentation and Structure in the Southern Uplands. In: "The British Caledonides", (Eds. M.R.W. Johnson & F.H. Stewart). Oliver and Boyd, Edinburgh and London.

Walton, E.K. 1965. Lower Palaeozoic rocks - stratigraphy. In: The Geology of Scotland, (Ed. G.Y. Craig). Oliver & Boyd, Edinburgh and and London.

Walton, E.K. 1965. Lower Palaeozoic rocks - palaeogeography and structure. In: The Geology of Scotland, (Ed. G.Y. Craig). Oliver and Boyd, Edinburgh and London.

Williams, A. 1962. The Barr and Lower Ardmillan Series (Caradoc) of the Girvan district, south-west Ayrshire, with a description of the brachiopods. Mem. Geol. Soc. Lond. No. 3.

Williams, A. 1969. Ordovician of British Isles. In: North Atlantic - Geology and Continental Drift. Mem. 12, Amer. Assoc. petrol. Geol. 236.

Williams, A. 1969. Ordovician Faunal Provinces with reference to Brachiopod Distribution. In: The Pre-Cambrian and Lower Palaeozoic Rocks of Wales. (Ed. A. Wood). Univ. of Wales Press, Cardiff.

Wilson, J.T. 1966. "Did the Atlantic close and Re-open again?" Nature, London. 211, 676.

Wood, A. (ed.) 1969. The Precambrian and Lower Palaeozoic rocks of Wales. Univ. of Wales Press, Cardiff.

Wood, A. & Smith, A.J. 1958. The sedimentation and sedimentary history
of the Aberystwyth Grits (Upper Llandoverian).
Quart.Journ. Geol. Soc. Lond. 114, 163.

Wright, A.E. 1969. Precambrian Rocks of England, Wales and Southeast
Ireland. In: North Atlantic - Geology and Continental Drift.
Mem. 12, Amer. Assoc. petrol. Geol. 93.

Ziegler, A.M. 1970. Geosynclinal development of the British Isles during
the Silurian Period. J. Geol. 78, 445.


Chapter 3

Anderson, J.G.C. 1965. The Precambrian of the British Isles. In:
The Geologic Systems: The Precambrian, (Ed. K. Rankama).
Vol. Wiley, New York and Chichester.

Briden, J.C. 1973. Applicability of Plate Tectonics to Pre-Mesozoic time.
Nature, London. 244, 400.

Dearnley, R. 1962. An outline of the Lewisian Complex of the Outer
Hebrides in relation to that of the Scottish mainland.
Quart. Journ. Geol. Soc. Lond. 118, 143.

Dearnley & Dunning, F. 1968. Metamorphosed and deformed pegmatites and.basic
dykes in the Lewisian Complex of the Outer Hebrides and their geological
significance. Quart. Journ. Geol. Soc. Lond. 123, 335.

Dewey, J.F. & Horsfield, B. 1970. Plate Tectonics, Orogeny and
Continental Growth. Nature, London. 225, 521.

Fyfe, W.S. 1973. The Generation of Batholiths. Tectonophysics. 17, 273.

Fyfe, W.S. 1974. Archaean Tectonics. Nature, London. 249, 338.

Green, D.H. 1972. Magmatic activity as the major process in the chemical
evolution of the Earth's crust and mantle. Tectonophysics, 13, 47.

McGregor, V.R. 1973. The early Precambrian gneisses of the Godthab region.
Phil. Trans. Roy. Soc. Lond. A273, 343.

Ramsay, J.G. 1963. Structure and metamorphism of the Moine and Lewisian
rocks of the north-west Caledonides. In: The British Caledonides
(Eds. M.R.W. Johnson and F.H. Stewart). Oliver and Boyd, Edinburgh and
London.

Read, H.H. & Watson, J. 1975. Introduction to Geology, Vol. 2. Earth
History. Part I. Early Stages of Earth History. Macmillan Press Ltd.,
London and Basingstoke.

Rubey, W.W. 1951. Geological history of sea water: an attempt to slate
the problem. Bull. geol. Soc. Am. 62, 1111.

Sutton, J. 1963. Long-term cycles in the evolution of the continents.
Nature, London, 198, 731.

Sutton, J. 1967. The Extension of the Geological Record into the Precambrian.
Proc. Geol. Ass. Lond. 78, 493.

Sutton, J. & Watson, J. 1951. The pre-Torridonian metamorphic history of
the Loch Torridon and Scourie areas in the North-west Highlands and its
bearing on the chronological classification of the Lewisian.
Quart. Journ. Geol. Soc. Lond. 106, 241.

Watson, J.V. 1965. Lewisian. In: The Geology of Scotland, (Ed. G.Y. Craig).
Oliver and Boyd, Edinburgh and London.

White, A.J.R., Jakes, P. & Christie, D.M. 1971. Composition of Greenstones
and the hypothesis of Sea-Floor Spreading in the Archaean.
Spec. Publs. geol. Soc. Aust. 3, 47.

Windley, B.F. 1973. Crustal development in the Precambrian.
Phil. Trans. R. Soc. Lond. A.273, 321.

Windley, B.F. 1977. The crust-mantle boundary in space and time.
Jl. geol. Soc. Lond. 34, 99.

## Chapter 4

Bluck, B.J. 1969. Old Red Sandstone and Other Palaeozoic Conglomerates of Scotland. In: North Atlantic - Geology and Continental Drift. Mem. 12, Amer. Assoc. petrol. Geol. 711.

Bott, M.H.P. 1974. The geological interpretation of a gravity survey of the English Lake District and the Vale of Eden. Jl. geol. Soc. Lond. 130, 309.

Bromley, A.V. 1976. A new interpretation of the Lizard complex, south Cornwall in the light of the ocean crust model. Proc. geol. Soc. Lond. 132, 114.

Dineley, D.L. 1961. The Devonian System in South Devon. Field Studies. 1, 121.

George, T.N. 1965. The geological growth of Scotland. In: The Geology of Scotland, (Ed. G.Y. Craig). Oliver & Boyd, Edinburgh and London.

Hendriks, E.M.L. 1959. A summary of present views on the structure of Cornwall and Devon. Geol. Mag. 96, 253.

House, M.R. 1963. Devonian ammonoid successions and facies in Devon and Cornwall. Quart. Journ. Geol. Soc. London, 119, 1.

Kennedy, W.Q. 1946. The Great Glen Fault. Quart. Journ. Geol. Soc. Lond. 102, 41.

Mercy, E.L.P. 1965. Caledonian igneous activity. In: The Geology of Scotland, (Ed. G.Y. Craig). Oliver and Boyd, Edinburgh and London.

Walmsley, V.G. 1974. The Base of the Upper Palaeozoic. In: The Upper Palaeozoic and Post-Palaeozoic Rocks of Wales, (Ed. T. R. Owen). Univ. of Wales Press, Cardiff.

## Chapter 5

Butcher, N.E. 1962. The tectonic structure of the Malvern Hills. Proc. Geol. Ass. Lond. 73, 103.

Coe, K. (ed.). 1962. Some Aspects of the Variscan Fold Belt. 9th Inter-Univ. Cong. Manchester Univ. Press, Manchester.

Dixon, O.A. 1972. Lower Carboniferous rocks between the Curlew and Ose Mountains, north-west Ireland. Jl. Geol. Soc. Lond. 73, 111.

Dunham, K.C. & Poole, E.G. The Oxfordshire Coalfield. Jl. geol. Soc. Lond. 130, 387.

Dunham, K.C., Dunham, A.C., Hodge, B.L. & Johnstone, G.A.L. 1965. Granite beneath Visean sediments with mineralisation at Rookhope. Quart. Journ. Geol. Soc. Lond. 121, 283.

Francis, E.H. 1965. Carboniferous. In: The Geology of Scotland, (Ed. G.Y. Craig). Oliver and Boyd, Edinburgh and London.

Francis, E.H. 1965. Carboniferous-Permian Igneous Rocks. In: The Geology of Scotland, (Ed. G.Y. Craig). Oliver & Boyd, Edinburgh and London.

George, T.N. 1940. The structure of Gower. Quart. Journ. Geol. Soc. Lond. 96, 131.

George, T.N. 1958. Lower Carboniferous Palaeogeography of the British Isles. Proc. Yorks. Geol. Soc. 31, 227.

George, T.N. 1963. Palaeozoic growth of the British Caledonides. In: The British Caledonides, (Eds. Johnson, M.R.W. & Stewart, F.H.). Oliver and Boyd, Edinburgh and London.

Hendricks, E.M.L. 1959. A summary of present views on the structure of Cornwall and Devon. Geol. Mag. 96, 253.

Kelling, G. 1974. Upper Carboniferous sedimentation in South Wales.
  In: The Upper Palaeozoic and Post-Palaeozoic Rocks of Wales,
  (Ed. T. R. Owen). Univ. of Wales Press, Cardiff.
Lambert, J.L.M. 1965. A Reinterpretation of the Breccias in the Meneage
  crush zone of the Lizard Boundary, South-west England. Quart. Journ.
  Geol. Soc. Lond. 121, 339.
Mykera, W. 1951.    The age of the Malvern folding. Geol. Mag. 88, 386.
Phipps, C.B. & Reeve, F.A.E. 1969. Structural geology of the Malvern,
  Abberley and Ledbury hills. Quart. Journ. Geol. Soc. Lond. 125, 1.
Prentice, J.E. 1959. Dinantian, Namurian and Westphalian rocks of the
  district south-west of Barnstaple, North Devon. Quart. Journ. Geol.
  Soc. Lond. 115, 261.
Raw, F. 1952. Structure and origin of the Malvern Hills.
  Proc. Geol. Ass. Lond. 63, 227.
Reading, H.G. 1965. Recent finds in the Upper Carboniferous of Southwest
  England and their significance. Nature, Lond. 208, 745.
Russell, M.J. 1972. North-south geofractures in Scotland and Ireland.
  Scott. Journ. Geol. 8, 75.
Sadler, P.M. 1974. An appraisal of the "Lizard - Dodman - Start Thrust"
  concept. Proc. Ussher, Soc. 3, 1.
Selley, R. 1970. Ancient Sedimentary Environments. Chapman and Hall, Lond.
Trotter, F.M. 1948. The devolatilization of coal seams in South Wales.
  Quart. Journ. Geol. Soc. Lond.
Trueman, A.E. 1954. The Coalfields of Great Britain. Edward Arnold, Lond.
Vaughan, A. 1905. The palaeontological sequence in the Carboniferous
  Limestone of the Bristol area. Quart. Journ. Geol. Soc. Lond. 61, 181.
Wills, L.J. 1956. Concealed Coalfields. Blackie, London.
Woodland, A.W. & Evans, W.B. 1964. Geology of the South Wales Coalfield,
  Part IV. Pontypridd and Maesteg. Mem. geol. Surv. Gt. Britain. 3rd ed.

## Chapter 6

Hancock, J.M. 1961. The Cretaceous system in Northern Ireland.
  Quart. Journ. Geol. Soc. Lond. 117, 11.
Rose, G.N. & Kent, P.E. 1955. A Lingula bed in the Keuper of Nottinghamshire.
  Geol. Mag. 92, 476.
Selley, R.C. The Habitat of North Sea Oil. Proc. Geol. Ass. Lond.87, 359.
Shotten, F.W. 1956. Some aspects of the New Red Desert in Britain.
  L'pool. Manchr. geol. J. 1, 450.
Stewart, F.H. 1963. The Permian Lower Evaporites of Forden in Yorkshire.
  Proc. Yorks. Geol. Soc. 34, 1.
Walsh, P.T. 1966. Cretaceous outliers in south-west Ireland and their
  implications for Cretaceous palaeogeography. Quart. Journ. Geol. Soc.
  Lond. 122, 63.
Warrington, G. 1970. The stratigraphy and palaeontology of the 'Keuper'
  Series of the central Midlands of England. Quart. Journ. Geol. Soc. Lond.
  126, 183.
Wills, L.J. 1948. The palaeogeography of the Midlands. Hodder and
  Stoughton, London.
Wills, L.J. 1951. A palaeogeographical atlas of the British Isles and
  adjacent parts of Europe. Blackie, London and Glasgow.
Woodland, A.W. (Ed.) 1975. Petroleum and the Continental Shelf of North-West
  Europe. Vol. 1. Geology. Institute of Petroleum, Great Britain.
Wright, J.E., Hull, J.H., McQuillin, R. & Arnold, S.E. 1971. Irish Sea
  investigations, 1969-70. Rep. Inst. geol. Sci. 71/19.

Van der Voo, R. & French, R.B. 1974. Apparent Polar Wandering for the Atlantic-Bordering Continents: Late Carboniferous to Eocene. Earth Sc. Reviews. 10 (No. 2), 99.


Chapter 7

Bott, M.H.P. & Tuson, J. 1974. Interpretation of gravity surveys over the Tertiary volcanic centres of Skye, Mull, and Ardnamurchan. Rep. Inst. geol. Sci.

Brooks, M. & James, D.G. 1975. The geological results of seismic refraction surveys in the British Channel, 1970-1973. Jl. geol. Soc. Lond. 131, 163.

Coope, G.R., Morgan, A. & Osborne, P.J. 1971. Fossil Coleoptera as Indicators of Climatic Fluctuations during the Last Glaciation in Britain. Palaeogeog. Palaeoclim. Palaeoecol. 10, 87.

Daley, B. & Edwards, N. 1974. Week-end Field Meeting: The Upper Eocene-Lower Oligocene Beds of the Isle of Wight; Report by the Directors. Proc. Geol. Ass. Lond. 85, 281.

Dearman, W.R. 1963. Wrench faulting in Cornwall and Devon. Proc. Geol. Ass. Lond. 74, 265.

George, T.N. 1974. The Cenozoic Evolution of Wales. In: The Upper Palaeozoic and Post-Palaeozoic Rocks of Wales (Ed. T. R. Owen). Univ. of Wales Press, Cardiff.

Godwin, H. 1960. Radiocarbon dating and Quarternary history in Britain. Proc. R. Soc. 13, 153, 287.

Harland, W.B. 1969. Tectonic Evolution of North Atlantic Region. In: North Atlantic - Geology and Continental Drift. Mem. 12, Amer. Assoc. petrol. Geol. 817.

Hollingworth, S.E. 1938. The recognition and correlation of high-level Erosion Surfaces in Britain. Quart. Journ. Geol. Soc. Lond. 94, 55.

Jones, O.T. 1952. The drainage system of Wales and the adjacent regions. Quart. Journ. Geol. Soc. Lond. 107, 201.

Kidson, C. & Wood, R. 1974. The Pleistocene Stratigraphy of Barnstaple Bay. Proc. Geol. Ass. Lond. 85, 223.

Linton, D.L. 1951. Midland drainage: some considerations on its origin. Advmt. Sci. 7, 449.

Miller, J.A. & Fitch, F.J. 1962. Age of the Lundy granites. Nature, Lond. 195, 553.

Shotton, F.W. 1953. The Pleistocene deposits of the area between Coventry, Rugby and Leamington and their bearing upon the topographic development of the Midlands. Phil. Trans. R. Soc. Lond. B, 237, 209.

Shotton, F.W. 1967. The problems and contributions of methods of absolute dating within the Pleostocene period. Quart. Journ. Geol. Soc. Lond. 122, 357.

Stewart, F.H. 1965. Tertiary igneous activity. In: The Geology of Scotland, (Ed. G. Y. Craig). Oliver and Boyd, Edinburgh and London.

West, R.G. 1963. Problems of the British Quaternary. Proc. Geol. Ass. Lond. 74, 147.

West, R.G. 1967. The Quarternary of the British Isles. In: The Quarternary, (Ed. K. Rankama). Vol. 2, New York.

West, R.G. 1969. Pleistocene geology and biology. London.

Wooldridge, S.W. & Linton, D.L. 1955. Structure, surfaces and drainage in south-east England. London.

Wymer, J.J. 1974. Clactonian and Acheulian industries in Britain - their Chronology and Significance. Proc. Geol. Ass. Lond. 85, 931.

# Index